Chemical Approaches to the Synthesis of Inorganic Materials

Chemical Approaches to the Synthesis of Inorganic Materials

C.N.R. RAO, FRS

Solid State and Structural Chemistry Unit
Indian Institute of Science
and
Jawaharlal Nehru Centre for
Advanced Scientific Research
Bangalore - 560 012, India

(Also at the School of Chemistry,
University of Wales, Cardiff, UK)

JOHN WILEY & SONS
NEW YORK • CHICHESTER • BRISBANE • TORONTO • SINGAPORE

First Published in 1994 by
WILEY EASTERN LIMITED
4835/24 Ansari Road, Daryaganj
New Delhi 110 002, India

Distributors:

Australia and New Zealand:
JACARANDA WILEY LIMITED
PO Box 1226, Milton Old 4064, Australia

Canada:
JOHN WILEY & SONS LIMITED
Baffins Lane, Chichester, West Sussex, England

South East Asia:
JOHN WILEY & SONS (PTE) LIMITED
05-04, Block B, Union Industrial Building
37 Jalan Pemimpin, Singapore 2057

Africa and South Asia:
WILEY EASTERN LIMITED
4835/24 Ansari Road, Daryaganj
New Delhi 110 002, India

North and South America and rest of the World:
JOHN WILEY & SONS. INC.
605 Third Avenue, New York, NY 10158, USA

Copyright © 1994 WILEY EASTERN LIMITED
New Delhi, India

Library of Congress Cataloging-in-Publication Data

ISBN 0-470-23431-8 John Wiley & Sons
ISBN 81-224-0618-1 Wiley Eastern Limited

Typeset by Grover DTP Systems, Delhi and Printed at
Baba Bharkha Nath Printers, New Delhi, India

Dedicated to my
dear friends, colleagues and coworkers of
the solid state chemistry community

Preface

Chemical methods of synthesis play a crucial role in designing and discovering novel materials, specially metastable ones which cannot be prepared otherwise. They also provide better and less cumbersome methods of preparing known materials. The tendency nowadays is to avoid brute-force methods and instead employ methods involving mild reaction conditions. Soft chemistry routes are indeed becoming popular and will undoubtedly be pursued with greater vigour in the future. In view of the increasing importance being attached to materials synthesis, it was considered appropriate to outline the chemical methods of synthesis of inorganic materials in a small monograph and hence this effort.

In this monograph, we briefly examine the different types of reactions and methods employed in the synthesis of inorganic solid materials. Besides the traditional ceramic procedures, we discuss precursor methods, combustion method, topochemical reactions, intercalation reactions, ion-exchange reactions, alkali-flux method, sol-gel method, electrochemical methods, pyrosol process, arc and skull methods and high pressure methods. The last topic includes hydrothermal synthesis as well. Superconducting cuprates and intergrowth structures are discussed in separate sections. Synthetic methods for metal borides, carbides, nitrides, fluorides, silicides, phosphides and chalcongenides are also outlined. While it is not expected to serve as a laboratory guide, it is my hope that the monograph provides an up-to-date account of the varied aspects of chemical synthesis of inorganic materials and serves as a ready reckoner as well as an introduction to the subject to students, teachers and practitioners. The references cited in the monograph would help to obtain details regarding preparative procedures and related aspects.

June 30, 1993   **C.N.R. Rao**

Contents

1

Introduction

Much chemical ingenuity is involved in the synthesis of solid materials [1-6] and this aspect of materials science is getting increasingly recognized as a crucial component of the subject. Tailor-making materials of the desired structure and properties is the main goal of materials science and solid state chemistry, but it may not always be possible to do so. While one can evolve a rational approach to the synthesis of solid materials [7], there is always the element of serendipity, encountered not so uncommonly. A good example of an oxide discovered in this manner is $NaMo_4O_6$ (Fig. 1.1) containing condensed Mo_6 octahedral metal clusters [8]. This was discovered by Torardi and McCarley in their effort to prepare the lithium analogue of $NaZn_2Mo_3O_8$. Another chance discovery is that of the phosphorus-tungsten bronze, $Rb_xP_9W_{32}O_{112}$, formed by the reaction of phosphorus present in the silica of the ampoule, during the preparation of the Rb-WO_3 bronze [9]. Since the material could not be prepared in a platinum crucible, it was suspected that a constituent of the silica ampoule must have got incorporated. This discovery led to the synthesis of the family of phosphorus-tungsten bronzes of the type $A_xP_4O_8 (WO_3)_{2m}$. Chevrel compounds of the type $A_xMo_6S_8$ (A = Cu, Pb, La etc.) shown in Fig. 1.2 were also discovered accidentally [10].

Rational synthesis of materials requires a knowledge of crystal chemistry besides thermodynamics, phase equilibria and reaction kinetics. There are any number of examples of rational synthesis. A good example is of SIALON [11] where Al and oxygen were partly substituted for Si and nitrogen in Si_3N_4. The fast Na^+ ion conductor NASICON, $Na_3Zr_2PSi_2O_{12}$ (Fig. 1.3), was synthesized with a clear understanding of the coordination preferences of the cations and the nature of the oxide networks formed by them [12]. The zero-expansion ceramic $Ca_{0.5}Ti_2P_3O_{12}$ possessing the NASICON framework was later synthesized based on the idea that the property of zero-expansion would be exhibited by two or three coordination polyhedra linked in such a manner so as to leave substantial empty space in the network [7]. Synthesis of silicate-based porous materials, making use of organic templates to predetermine the pore or cage geometries, is well known [13].

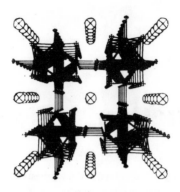

Fig 1.1 Structure of $NaMo_4O_6$ (after Torardi and McCarley, 1979)

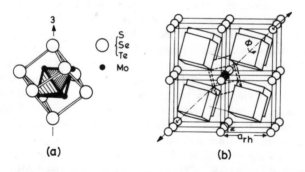

Fig. 1.2 Cheverl compounds, $A_xMo_6Ch_8$ (Ch = S, Se, Te)
(a) Mo_6Ch_8 Building block, (b) structure of Chevrel compounds where the
A cation is shown by the closed circle

A microporous phosphate of the formula $(Me_4N)_{1.3}$ $(H_3O)_{0.7}$ Mo_4O_8 $(PO_4)_2.2H_2O$ where the tretramethyl-ammonium ions fill the voids in the 3-dimensional structure made up of Mo_4O_8 cubes and PO_4 tetrahedra, has been prepared in this manner [14].

A variety of inorganic solids have been prepared in the past several years by the traditional ceramic method, which involves mixing and grinding powders of the constituent oxides, carbonates and such compounds and heating them at high temperatures with intermediate grinding when necessary. A wide range of conditions, often bordering on the extreme, such as high temperatures and pressures, very low oxygen fugacities and rapid

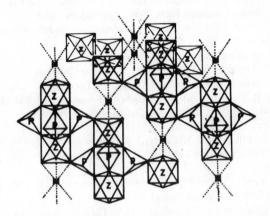

Fig. 1.3 Structure of $NaZr_2(PO_4)_3$ which provided the design for NASICON: vacant trigonal-prismatic sites, p; octahedral Zr^{4+} sites, Z and octahedral sites available for Na^+, M. For each M, There are three M_o sites forming hcp layers perpendicular to the c-axis (After J.B. Goodenough, *Proc. R. Soc. Lond.* 1984).

quenching have all been employed in materials synthesis. Low-temperature chemical routes and such methods involving mild reaction conditions are, however, of greater interest. The present-day trend is to avoid brute-force methods in order to get a better control of the structure, stoichiometry and phasic purity. The so-called *soft- chemistry* routes, which the French call *chimie douce*, are indeed desirable because they lead to novel products, many of which are metastable and cannot otherwise be prepared. Soft chemistry routes essentially make use of simple reactions such as intercalation, ion exchange, hydrolysis, dehydration and reduction that can be carried out at relatively low temperatures. The topochemical nature of certain solid state reactions is also exploited for synthesis. Ion exchange, intercalation and many other types of reactions are generally topochemical.

In the sections that follow, we shall briefly discuss the synthesis of inorganic solids by various methods with several examples, paying greater attention to the chemical routes. While oxide materials occupy a greater part of the monograph, other classes of materials such as chalcogenides, carbides, fluorides and nitrides have been discussed. Superconducting cuprates and intergrowth structures have been discussed in separate sections.

References

1. P. Hagenmuller (ed) *Preparative Methods in Solid State Chemistry.* Academic Press, New York, 1972.

2. C.N.R. Rao and J. Gopalakrishnan, *New Directions in Solid State Chemistry*, Cambridge University Press, 1986, paperback ed. 1989

3. A.R. West, *Solid State Chemistry and Applications*. John Wiley, Chichester, 1984.

4. F.J. Di Salvo, *Science. 247* (1990) 647.

5. A.K. Cheetham and P. Day (eds) *Solid State Chemistry —Techniques and Compounds*, Clarendon Press, Oxford, 1987 and 1992.

6. C.N.R. Rao (ed) *Chemistry of Advanced Materials* (IUPAC 21st Century Monograph Series) Blackwell, Oxford, 1992.

7. R. Roy, *Solid State Ionics, 32-33* (1989) 3.

8. C. Torardi and R.E. McCarley, *J. Am. Chem. Soc., 101* (1979) 3963.

9. J.P. Giroult, M. Goreaud, P.H. Labbe and B. Raveau, *Acta Cryst., B36* (1980) 2570.

10. K. Yvon, *Current Topics in Materials Science*, Vol. 3 (E. Kaldis, ed), North-Holland, Amsterdam, 1979.

11. K.H. Jack, *Mater. Res. Bull.*, 13 (1978) 1327.

12. J.B. Goodenough, H.Y.P. Hong and J.A. Kafalas, *Mater. Res. Bull., 11* (1976) 203.

13. J.M. Newsam in *Solid State Chemistry-Compounds* (A.K. Cheetham and P. Day, eds), Clarendon Press, Oxford, 1992.

14. R.C. Haushalter, K.G. Storhmaeu and F.W. Lai, *Science 246* (1989) 1289.

2

Common Reactions Employed in Synthesis

Various types of chemical reactions have been used for the synthesis of solid materials [1, 2]. Corbett [1] has written a fine article on the subject. Some of the common reactions employed for the synthesis of inorganic solids are:

1. **Decomposition**

   $$A\ (s) \longrightarrow B\ (s) + C\ (g)$$
   $$A\ (g) \longrightarrow B\ (s) + C\ (g)$$

2. **Addition**

   $$A\ (s) + B\ (g) \longrightarrow C\ (s)$$
   $$A\ (s) + B\ (s) \longrightarrow C\ (s)$$
   $$A\ (s) + B\ (1) \longrightarrow C\ (s)$$
   $$A\ (g) + B\ (g) \longrightarrow C\ (s)$$

3. **Metathetic reaction (which combines 1 and 2 above)**

   $$A\ (s) + B\ (g) \longrightarrow C\ (s) + D\ (g)$$

4. **Other exchange reactions**

   $$AX\ (s) + BY\ (s) \longrightarrow AY\ (s) + BX\ (s)$$
   $$AX\ (s) + BY\ (g) \longrightarrow AY\ (s) + BX\ (g)$$
   $$AX\ (s) + BY\ (1) \longrightarrow AY\ (s) + BX\ (s)$$

Typical examples of the above reactions are:

1. $CaCO_3 (s) \longrightarrow CaO (s) + CO_2 (g)$

$M_mO_n(s) \longrightarrow M_mO_{n-\delta} (s) + \dfrac{\delta}{2} O_2(g)$ (M = metal)

$SiH_4(g) \longrightarrow Si(s) + 2H_2(g)$

2    $YBa_2Cu_3O_6(s) + O_2(g) \longrightarrow YBa_2Cu_3O_7(s)$

$ZnO(s) + Fe_2O_3(s) \longrightarrow ZnFe_2O_4(s)$

$BaO(s) + TiO_2(s) \longrightarrow BaTiO_3(s)$

$Cd(s,1) + CdX_2(s,1) \longrightarrow Cd_2X_2(s)$

3. $CO(g) + MnO_2(s) \longrightarrow CO_2(g) + MnO(s)$

$Pr_6O_{11}(s) + 2H_2(g) \longrightarrow 3Pr_2O_3(s) + 2H_2O(g)$

$MO_n(s) + nH_2S(g) \longrightarrow MS(s) + nH_2O(g)$ (M = Metal)

$SiCl_4(s)\ 2H_2(g) \longrightarrow Si(s) + 4HCl(g)$

$3\ SiCl_4(g) + 4NH_3(g) \longrightarrow Si_3N_4(s) + 12HCl(g)$

$GaMe_3(g) + AsH_3(g) \longrightarrow GaAs(s) + 3\ CH_4(g)$

$MR_x (g)\ +\ M'R'_y(g) \longrightarrow MM'(s)\ +\ R_xR'_y(g)$
$$(R = alkyl, M = Cd, Te\ etc.)$$

4. $ZnS(s) + CdO(s) \longrightarrow CdS(s) + ZnO(s)$

$MnCl_2(s) + 2HBr \longrightarrow MnBr_2(s) + 2HCl$

$LiFeO_2(s) + CuCl(1) \longrightarrow CuFeO_2(s) + LiCl(s)$

Complex reactions involving more than one type of reaction are also commonly employed in solid state synthesis. For example, in the preparation of complex oxides, it is common to carry out thermal decomposition of a compound followed by oxidation (in air or O_2) essentially in one step.

$$2Ca_{0.5}Mn_{0.5}CO_3(s) + \dfrac{1}{2} O_2(g) \longrightarrow Ca\ Mn\ O_3(s) + 2CO_2(g)$$

Vapour phase reactions and liquid-gas reactions yield solid products in many instances. For example, the reaction of $TiCl_4$ and H_2S gives solid TiS_2 and HCl gas. Reaction of metal halides with NH_3 to yield nitrides is another example.

In chemical vapour transport reactions, a gaseous reagent acts as a carrier to transport a solid by transforming it into the vapour state. For example, $MgCr_2O_4$ cannot be readily formed by the reaction of MgO and Cr_2O_3. However, Cr_2O_3 (s) reacts with O_2 giving CrO_3 (g) which then reacts with MgO giving the chromate. The overall reaction is,

$$MgO(s) + Cr_2O_3(s) \xrightarrow{O_2} MgCr_2O_4(s)$$

Some of the typical transport reaction equilibria are:

$$ZnS(s) + I_2 \rightleftharpoons ZnI_2 + \frac{1}{2} S_2$$

$$TaOCl_2(s) + TaCl_5 \rightleftharpoons TaOCl_3 + TaCl_4$$

$$Nb_2O_5(s) + 3\, NbCl_5 \rightleftharpoons 5\, NbOCl_3$$

$$GaAs(s) + HCl \rightleftharpoons GaCl + \frac{1}{2} H_2 + As$$

Transport of two substances in opposite directions is possible if the reactions have opposite heats of reaction. For example, Cu_2O and Cu can be separated by using HCl as the transporting agent.

$$Cu_2O(s) + 2HCl(g) \underset{1070\,K}{\overset{770\,K}{\rightleftharpoons}} 2CuCl(g) + H_2O(g)$$

$$Cu(s) + HCl(g) \underset{770\,K}{\overset{870\,K}{\rightleftharpoons}} CuCl(g) + \frac{1}{2} H_2(g)$$

Another example of this kind is the separation of WO_2 and W by using I_2 (g), involving the formation of WO_2I_2 (g). Volatility of the product also allows its separation from other species. Thus, the reaction of Cl_2 gas with a solid mixture of Al_2O_3 and carbon yields $AlCl_3$ and CO gas.

Vapour transport methods are used in the synthesis of materials as exemplified by the reaction of MgO and Cr_2O_3; another example is the formation of $NiCr_2O_4$ involving the CrO_3 (g) species

$$Cr_2O_3(s) + \frac{3}{2}O_2(g) \rightleftharpoons 2\,CrO_3(g)$$

$$2\,CrO_3(g) + NiO(s) \longrightarrow NiCr_2O_4(s) + \frac{3}{2} O_2(g)$$

The formation of Ca_2SnO_4 by the reaction of CaO and SnO_2 is facilitated by CO *via* the formation of gaseous SnO which then reacts with CaO. Zn WO_4 is made by heating ZnO and WO_3 at 1330 K in the presence of Cl_2 gas (volatile chlorides being the intermediates). In the reaction of Al and sulfur to form Al_2S_3 by using I_2, the sulfide is transported through the formation of AlI_3

$$2\,Al + 3S \longrightarrow Al_2S_3$$

$$Al_2S_3(s) + 3I_2(g) \rightleftharpoons 2AlI_3(g) + \frac{3}{2}\,S_2(g)$$

Cu_3TaSe_4 if formed by the reaction of Cu, Ta and Se in the presence of gaseous I_2. In Table 2.1 we list a few examples of chemical transport systems. Table 2.2 lists some crystals grown by the chemical vapour transport method.

Table 2.1

Examples of chemical transport

Solid	Transporting Agent	Solid	Transporting Agent
Nb_2O_5	Cl_2, $NbCl_5$	CrOCl	Cl_2
TiO_2	$I_2 + S_2$	$FeWO_4$	Cl_2
IrO_2	O_2	$MgFe_2O_4$	HCl
WO_3	H_2O	$CaNb_2O_6$	Cl_2, HCl
NbS_2	S	ZrOS	I_2
TaS_3	S	$LaTe_2$	I_2
$MnGeO_3$	HCl	V_nO_{2n-1}	$TeCl_4$
$MgTiO_3$	Cl_2	NbS_2Cl_2	$NbCl_4$

Table 2.2

Examples of crystals grown by the chemical vapour transport method

Starting materials	Product (crystal grown)	Transport agent	$T\ (K)$
SiO_2	SiO_2	HF	470 – 770
Fe_3O_4	Fe_3O_4	HCl	1270 – 1070
Cr_2O_3	Cr_2O_3	$Cl_2 + O_2$	1070 – 870
$MO+Fe_2O_3$	MFe_2O_4	HCl	–
(M = Mg, Co, Ni)			
$Nb+NbO_2$	NbO	Cl_2	–
$NbSe_2$	$NbSe_2$	I_2	1100 – 1050

Oxidation of many metals occurs slowly. Thus, oxidation of Cu stops at the stage of Cu_2O at 1270 K in oxygen. In order to promote further oxidation (e.g. to CuO in the case of Cu), an easily oxidizable salt is used (e.g. CuI

$\longrightarrow$ CuO at 620 K). Similarly, fluorination of a compound may be easier than that of the native metal (e.g. $CuCl_2 \longrightarrow CuF_2$ in the presence of F_2, instead of $Cu + F_2$).

Reduction of oxides is carried out in an atmosphere of (flowing) pure or dilute hydrogen (e.g. N_2–H_2 mixtures) or sometimes in an atmosphere of CO or CO–CO_2 mixtures. Reduction of oxides for the purpose of lowering the oxygen content is also achieved by heating oxides in argon or nitrogen or by using other metals as getters (e.g. Ti or Zr sponge, molten Na) to remove some of the oxygen. Thus, the oxygen content of $YBa_2Cu_3O_{7-\delta}$ can be varied by heating in N_2 or by heating in presence of hot Ti sponge. Application of vacuum at an appropriate temperature (vacuum annealing or decomposition at low pressures) is also used. Exact control of oxygen stoichiometry in oxides such as Fe_3O_4 or V_2O_3 is accomplished by annealing the oxide in CO-CO_2 mixtures of known oxygen fugacity at an appropriate temperature. In preparing oxides of exact stoichiometry, it is necessary to have the fugacity diagrams of the type shown in Fig. 2.1.

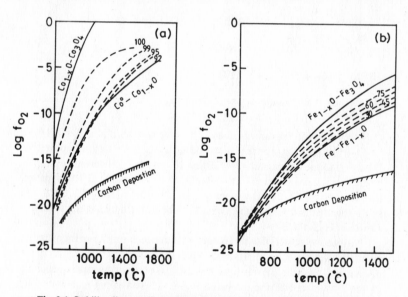

Fig. 2.1 Stability diagrams for (a) $Co_{1-x}O$ and (b) $Fe_{1-x}O$ in $\log f(O_2)$-temperature representation. Upper solid line gives the oxidation limit and the lower solid line the reduction limit. Dashed lines, CO/CO_2 gas mixtures with percentage of CO_2 shown in number (i.e., $100CO_2/C$) + CO_2). (From Harrison et al., *Mat. Res. Bull.*, 1980)

The obvious means of reducing solid compounds is by hydrogen. Hydrogen reduction is employed not only for reducing oxides, but also halides and other compounds. Thermal decomposition of metal halides often yields lower halides.

$$M_2O_3(s) + H_2(g) \longrightarrow 2MO(s) + H_2O(g) \quad (e.g. \ M = Fe)$$

$$ABO_3(s) + H_2(g) \longrightarrow ABO_{2.5}(s) + \frac{1}{2}H_2O(g) \ (e.g. \ LaCoO_3 \ and \ CaMnO_3)$$

$$MCl_3(s) + H_2(g) \longrightarrow MCl_2(s) + HCl(g) \quad (e.g. \ M = Fe, Cr)$$

$$MCl_2(s) + H_2(g) \longrightarrow M(s) + 2HCl(g) \quad (e.g. \ M = Cr)$$

$$MX_3 \xrightarrow{\text{heat}} MX_2 + \frac{1}{2}X_2 \quad (e.g. \ M = Cr)$$

Reduction of oxides can be accomplished by reacting with elemental carbon or with a metal. Reduction of halides is also carried out by metals.

$$2MCl_3 + M \longrightarrow 3MCl_2 \quad (e.g. \ M = Nd, Fe)$$

$$3MCl_4 + M'(s) \longrightarrow 3MCl_3(s) + M'Cl_3(g)$$

$$Nb_2O_5 + 3Nb \longrightarrow 5NbO$$

$$TiO_2 + Ti \longrightarrow 2TiO$$

Metals such as aluminium are used as reducing agents for other metal halides.

$$3HfCl_4 + Al \longrightarrow 3HfCl_3 + AlCl_3$$

Metal oxychlorides are obtained by heating oxides with Cl_2 (LaOCl from La_2O_3). Fluorination is generally carried out by using elemental fluorine, HF or other fluorine compounds (see Section 18 for details). There are examples where oxides are reacted with a fluoride such as BaF_2 to attain partial fluorination. Sulfidation is generally carried out by heating the metal and sulfur together in a sealed tube. Oxides can be sulfided by heating them in a stream of H_2S or CS_2.

Plasma or electrical discharge reactions have been employed for materials synthesis. Amorphous silicon is produced by the decomposition of SiH_4 under discharge. Unusual compounds such as $ZrCl_3$ is obtained by rapid quenching of the plasma out of the discharge region. Plasma spray techniques are employed to prepare films of materials. In the presence of oxygen, the plasma technique is useful in preparing certain oxides as exemplified by oxygen-excess La_2CuO_4.

Substitution of one metal ion by another is often carried out to attain new structures and properties. For example, partial substitution Ni in metallic $LaNiO_3$ by Mn makes it non-metallic. On the other hand, partial substitution of Ln^{3+} by Sr^{2+} in insulating $LnCoO_3$ (Ln = La, Pr, Nd etc.) makes the d-electrons itinerant and the material becomes ferromagnetic. Thus, $La_{0.5}$ $Sr_{0.5}CoO_3$ is a ferromagnetic metal [3]. Similar changes are brought about by the substitution of La^{3+} by Sr^{2+} or Ca^{2+} in $LaMnO_3$ [4]. Partial substitution of V by Ti in V_2O_3 wipes out the metal insulator transition and makes the material metallic. In the non-linear optical material, $KTiOPO_4$, tetravalent Ti can be usefully replaced partly by pentavalent Nb, provided P is proportionately replaced by Si as in $KTi_{0.5}Nb_{0.5}OP_{0.5}$ $Si_{0.5}O_4$ [5]. Relative ionic size and charge neutrality, by and large, govern these substitutions.

Soft chemistry routes: It was pointed out earlier that *soft chemistry routes* are receiving considerable attention recently. It would be instructive to examine a few typical examples of soft chemical methods of synthesis of materials (*chimie douce*). Marchand et al [6] obtained a new form of TiO_2 by the dehydration of H_2TiO_9. xH_2O, which in turn was prepared by the exchange of K^+ by H^+ in $K_2Ti_4O_9$. The mechanism of this transformation has been described recently by Feist and Davis [7] and we show this schematically in Fig. 2.2. Rebbah et al [8] prepared $Ti_2Nb_2O_9$ by the dehydration of $HTiNbO_5$, the latter having been prepared from $KTiNbO_5$ by cation exchange (Fig. 2.3). A fine example that typifies an entire class of reactions yielding novel, metastable materials is the oxidative deintercalation of $LiVS_2$ to give VS_2 which cannot otherwise be prepared [9]. A new form of FeF_3 was prepared by De Pape and Ferey [10] by the topotactic oxidation of $NH_4Fe_2F_6$ by Br_2 in acetonitrile (giving $(NH_3)_xFeF_3$ along with HBr and NH_4Br), which followed by heating at 480 K in vacuum gives pyrochlore type FeF_3. Delmas et al. prepared $Ni(OH)_2$. xH_2O with a large intersheet distance of 7.8 Å by the hydrolysis of $NaNiO_2$ to $NiOOH$ followed by reduction. Delmas and coworkers have also prepared layered double

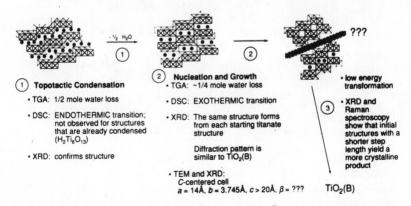

Fig. 2.2 Mechanism of formation of metastable TiO_2 (B) from $K_2Ti_4O_9$ (From Feist and Davis, *J. Solid State Chem.* 1992).

hydroxides of the type $Ni_{1-x}M_x(OH)_2X^{n-}\cdot zH_2O$ (M = Co or Fe) starting from $NaNi_{1-x}M_xO_2$ [11, 12]. We show the transformations involved in Fig. 2.4.

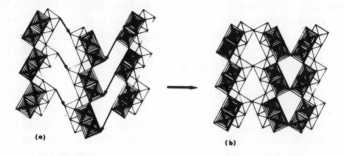

Fig. 2.3 Preparation of $Ti_2Nb_2O_8$ (b) from $KTiNbO_5$ (a) (From Rebbah et al, *Mat. Res. Bull.* 1979).

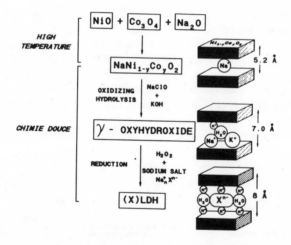

Fig. 2.4 Preparation of layered double hydroxides (LDH). The thickness of the $Ni_{1-y}Co_yO_2$ slab varies with the oxidation state of nickel and cobalt (From Delmas and Borthomieu, *J. Solid State Chem.* 1993).

Many of the chemical methods such as electrochemical oxidation and intercalation are also soft-chemical routes. The synthesis of $HAlO_2$ from α-$LiAlO_2$ is an interesting example of soft chemistry [13]. This proton-stuffed alumina is prepared by the reaction of $LiAlO_2$ with molten lauric or benzoic acid. Acid-base chemistry of α-Zr $(HPO_4)_2H_2O$ (α-ZrP),

HSb $(PO_4)_2$ and H_3Sb3O_6 $(PO_4)_2$, their exchange properties in acidic medium and intercalation properties involve soft chemical reactions [14]. Hydrated acids of the type $H_4Sb_4O_8Si_4O_{12}$ 6.5 H_2O and $H_3Sb_3O_6$ (Si_2O_7) $6H_2O$ have been prepared starting from Cs_2O– Sb_2O_5–SiO_2 [15]. Fibers of $K_2Ti_6O_{13}$ are excellent materials for heat insulation and resistance, but cannot be prepared by the reaction of TiO_2 with K_2O. They are obtained starting from fibrous crystals of $K_2Ti_4O_9$ (prepared by flux growth) by replacing K^+ partially by H^+ through ion exchange, followed by heating [16]. Unlike crystalline lamellar thiophosphate, MPS_3 (M = V, Mn, Fe etc.) which are prepared by the high temperature reaction between the elements, amorphous thiophosphates can be prepared by the reaction of Li_2PS_3 with a transition metal salt [17]. Topochemical dehydration, oxidation ion exchange and other reactions employed to prepare many oxides can also be considered to be soft chemistry routes. For example, metastable MoO_3 in the ReO_3–like structure can be prepared by the slow dehydration of $MoO_3.H_2O$ [18] or by the oxidation of Mo_4O_{11} [19]. We shall be discussing a variety of chemical methods of preparing novel materials in the subsequent sections.

References

1. J.D. Corbett in *Solid State Chemistry–Techniques* (A.K. Cheetham and P. Day, eds), Clarendon Press, Oxford. 1987
2. C.N.R. Rao and J. Gopalakrishnan, *New Directions in Solid State Chemistry*, Cambridge University Press. 1989
3. C.N.R Rao, D. Bahadur, O. Parkash and P. Ganguly, *J. Solid State Chem. 22* (1977) 353.
4. J.B. Goodenough. *Progress in Solid State Chemistry, 5* (1971) 149.
5. K. Kasturi Rangan, B.R. Prasad, C.K. Subramanian and J. Gopalakrishnan, *Inorg. Chem. 32* (1993) 4291.
6. R. Marchand, L. Brohan and M. Tournoux, *Mater. Res. Bull., 15,* (1980) 1129.
7. T.P. Feist and P.K. Davis, *J. Solid State Chem., 101* (1992) 275.
8. H. Rebbah, G. Desgardin and B. Raveau, Mater. Res. Bull., 14 (1979) 1131.
9. D.W. Murphy, C. Cros, F.J. DiSalvo and J.V. Waszezak, *Inorg. Chem., 16* (1977) 3027.
10. B. De Pape and G. Ferey, *Mater. Res. Bull., 21* (1986) 971.
11. C. Delmas and Y. Borthomieu, *J. Solid State Chem., 104* (1993) 345.
12. L.D. Guerlou, J.J. Braconnier and C. Delmas, *J. Solid State. Chem., 104* (1993) 359.
13. J.P. Thiel, C.K. Chiang and K.R. Poeppelmeier, *Chem. Mater,* 5 (1992) 297; also see *Catal. Lett., 12* (1992) 139.
14. Y. Piffard, A Verbaere, A. Lachgar, S. Deniard-Courant and M. Tournoux, *Eur. J. Solid State Chem. 26* (1989) 113; also see *Eur. J. Solid State Chem. 26* (1989) 175.

15. C. Pagnouz, A. Verbaere, Y. Piffard and M. Tournoux, *Eur. J. Solid State chem. 30* (1983) 111.
16. Y. Fujiki, *JSAE Rev.* Nov. 91 (1981). Also see M. Watanabe in Proceedings of Intnl. Symposium on Soft Chemistry Routes to New Materials, Nantes, France, 1993 (Trans Tech Publications).
17. E. Prouzet, G. Ouvrard, R. Brec and P. Seguineau, *Solid State Ionics, 31* (1988) 79. Also G. Ouvard, E. Prouzet, R. Brec and J. Rouxel, To be published in the Proceedings of the Intnl, Symposium on Soft Chemistry Routes to New Materials, Nantes, France, 1993 (Trans Tech Publications).
18. C.N.R. Rao, J. Gopalakrishnan, K. Vidyasagar, A.K. Ganguli, A. Ramanan and L. Ganapathi, *J. Mat. Res., 1,* (1986) 280.
19. L. Kihlborg and coworkers, *Reactivity of Solids, 3* (1987) 33.

3

Ceramic procedures

The most common method of preparing metal oxides and other solid materials is by the ceramic method which involves grinding powders of oxides, carbonates, oxalates or other compounds containing the relevant metals and heating the mixture at a desired temperature, generally after pelletizing the material. Several oxides, sulfides, phosphides, etc., have been prepared by this method. A knowledge of the phase diagram is generally helpful in fixing the desired composition and conditions for synthesis. Some caution is necessary in deciding the choice of the container. Platinum, silica and alumina containers are generally used for the synthesis of metal oxides, while graphite containers are employed for sulfides and other chalcogenides as well as pnictides. Tungsten and tantalum containers are quite inert to metals and halides and have been used in many preparations, specially of halides. If one of the constituents is volatile or sensitive to the atmosphere, the reaction is carried out in sealed evacuated capsules. Most ceramic preparations require relatively high temperatures which are generally attained by resistance heating. Electric arc and skull techniques give temperatures up to 3300K while high power CO_2 lasers give temperatures up to 4300K.

The ceramic method suffers from several disadvantages. When no melt is formed during the reaction, the entire reaction has to occur in the solid state, initially by a phase boundary reaction at the points of contact between the components and later by the diffusion of the constituents through the product phase. With the progress of the reaction, diffusion paths become increasingly longer and the reaction rate slower. The product interface between the reacting particles acts as a barrier. The reaction can be speeded up to some extent by intermittent grinding between heating cycles. There is no simple way of monitoring the progress of the reaction in the ceramic method. It is only by trial and error (by carrying out x-ray diffraction and other measurements periodically) that one decides on appropriate conditions that lead to the completion of the reaction. Because of this difficulty, one frequently ends up with mixtures of reactants and products. Separation of the desired product from such mixtures is generally difficult, if not impossible. It is sometimes difficult to obtain a compositionally homogeneous product by the ceramic technique even where the reaction proceeds almost to completion.

In spite of such limitations, ceramic techniques have been successfully used for the synthesis of a variety of solid materials. Cation substitutions referred to in the previous section have been routinely carried out in many oxide systems (e.g. $La_{1-x}M_xBO_3$ where M = Ca, Sr and B = V, Mn or Co, $LaMM'O_3$ where M, M' = Mn, Fe, Co or Ni and $LnBa_2Cu_3O_7$ when Ln = Y, Pr, Nd, Gd etc.) by the ceramic method. Mention must be made, among others, of the use of this technique for the synthesis of rare earth mono-chalcogenides such as SmS and SmSe. The method involves heating the elements, first at lower temperatures (870-1170K) in evacuated silica tubes; the contents are then homogenized, sealed in tantalum tubes and heated around 2300K by passing a high current through the tube [1].

Various modifications of the ceramic technique have been employed to overcome some of the limitations. One of the modifications relates to decreasing the diffusion path lengths. In a polycrystalline mixture of reactants, the individual particles are approximately 10 μm in size, which represents diffusion distances of roughly 10000 cells. By using freeze-drying, spray-drying, coprecipitation, and sol-gel and other techniques, it is possible to bring down the particle size to a few hundred angstroms and thus effect a more intimate mixing of the reactants. In *spray-drying*, Suitable constituents dissolved in a solvent are sprayed in the form of fine droplets into a hot chamber. The solvent evaporates instantaneously leaving behind an intimate mixture of reactants, which on heating at elevated temperatures gives the product. In *freeze-drying*, the reactants, in a common solvent, are frozen by immersing in liquid nitrogen and the solvent removed at low pressures.

In *coprecipitation*, the required metal cations, taken as soluble salts (e.g. nitrates), are coprecipitated from a common medium, usually as hydroxides, carbonates, oxalates, formates or citrates. In actual practice, one takes oxides or carbonates of the relevant metals, digests them with an acid (usually HNO_3) and to the solution so obtained, the precipitating reagent is added. It is important that the solid precipitating out is really insoluble in the mother liquor. The precipitate after drying is heated to the required temperature in a desired atmosphere to get the final product. Many of the superconducting cuprates have been prepared by the coprecipitation method [2]. For example, tetraethylammonium oxalate has been used to obtain a precipitate which on decomposition gives superconducting $YBa_2Cu_4O_8$. The decomposition temperature of such precipitates is generally lower than that of crystalline carbonates, oxalates etc. Homogeneous precipitation can yield crystalline or amorphous products. If all the relevant metal ions do not form really insoluble precipitates, it becomes difficult to control the stoichiometry.

References

1. A. Jayaraman, P.D. Dernier and L.D. Longinotti, *High Temp. High Press* 7 (1975) 1.
2. C.N.R. Rao, R. Nagarajan and R. Vijayaraghavan, *Supercond Sci. Tech.* 6 (1993) 1.

4

Precursor methods

It was pointed out earlier that diffusion distances for the reacting cations are rather large in the ceramic method. Diffusion distances are markedly reduced to a few angstroms by incorporating the cations in the same solid precursor (Fig. 4.1). Synthesis of complex oxides by the decomposition of precursor compounds has been known for sometime. For example, thermal decomposition of $LaCo(CN)_6 . 5H_2O$ and $LaFe(CN)_6 . 6H_2O$ in air readily yield $LaCoO_3$ and $LaFeO_3$ respectively. $BaTiO_3$ can be prepared by the

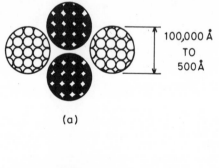

(a)

100,000 Å
TO
500Å

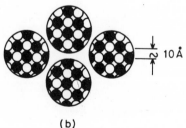

(b)

10Å

Fig. 4.1 Distribution of two different cations (closed and open circles) in reactant particles and the diffusion distances in (a) the ceramic procedure and (b) in precursor compounds or precursor solid solutions (Following Longo et al. 1980)

thermal decomposition of $Ba[TiO(C_2O_4)_2]$ while $LiCrO_2$ can be prepared from the hydrate of $Li[Cr(C_2O_4)_2]$. Ferrite spinels of the general formula MFe_2O_4 (M = Mg, Mn, Ni, Co) are prepared by the thermal decomposition of acetate precursors of the type $M_3Fe_6(CH_3COO)_{17}O_3$ $OH.12C_5H_5N$. Chromites of the type MCr_2O_4 are obtained by the decomposition of $(NH_4)_2$ $M(CrO_4)_2.6H_2O$.

In general, alkoxides and carboxylates are the precursors employed in the synthesis of metal oxides. Recently Chandler et al [1] have reviewed organometallic precursors employed in the synthesis of some perovskite oxides. It would be instructive to examine a few of the reactions involved in the synthesis of oxides from precursor compounds. Let us first examine the mode of formation of $BaTiO_3$ from the decomposition of barium titanyloxalate which is best represented as $Ba_2Ti_2(O)_2(C_2O_4)_4$.

$$Ba_2Ti_2(O)_2 (C_2O_4)_4 \longrightarrow Ba_2Ti_2(O)_2 (C_2O_4)_3 CO_3 + CO$$

$$Ba_2Ti_2(O)_2(C_2O_4) CO_3 \longrightarrow Ba_2Ti_2(O)_5 (CO_2) CO_3 + 2CO_2 + 3CO$$

$$Ba_2Ti_2(O)_5(CO_2) CO_3 \longrightarrow Ba_2Ti_2 (O)_5 (CO_3) + CO_2$$

$$Ba_2Ti_2(O)_5 CO_3 \longrightarrow 2BaTiO_3 + CO_2$$

$PbTiO_3$ can be prepared by making use of carboxylate and alkoxide precursors as follows:

$$PB(OAc)_2\ 3H_2O + CH_3OCH_2OH \xrightarrow[\text{distill}]{\text{reflux}} \text{Pb precursor}$$
$$\text{(after dehydration)}$$

$$Ti (O - i - Pr)_4 + CH_3OCH_2CH_2OH \xrightarrow[\text{distill}]{\text{reflux}} \text{Ti precursor}$$

By refluxing a 1 : 1 mixture of the Pb and Ti Precursors in the alcohol, we obtain the precursor for $PbTiO_3$. The precursor on decomposition gives $PbTiO_3$.

In order to prepare $PbZr_{1-x}Ti_xO_3$ and such oxides, the following procedure can be employed:

$$ACO_3 + 2HO_2CCR_2OH \xrightarrow{H_2O} A(O_2CCR_2OH)_2 + H_2O + CO_2$$
$$(A = Ca, Sr, Ba, Pb)$$

$$A(O_2CCR_2O)_2\ B(OR')_4 \longrightarrow A(O_2CCR_2O)_2 B (OR')_2 + 2R' OH$$
$$(B = Ti, Zr, Sn)$$

$$A (O_2CCR_2O)_2 B(OR')_2 \longrightarrow ABO_3$$

By taking appropriate mixtures of $A(O_2CCR_2O)_2$ $B(OR')_2$ with two different B′ cations, one can obtain $AB'_{1-x}B_xO_3$ type oxides.

Hydrazinate precursors have been employed to prepare a variety of oxides [2]. Metal-ceramic composites such as Fe/Al_2O_3 have been prepared by the thermal decomposition of complex ammonium oxalate precursors, $(NH_4)_3[Al_{1-x}Fe_x(C_2O_4)_3]\,nH_2O$ [3]. Organoaluminium silicate precursors have been employed to prepare aluminosilicates [4].

Carbonates of metals such as Ca, Mg, Mn, Fe, Co, Zn and Cd are all isostructural, possessing the calcite structure. We can, therefore, prepare a large number of carbonate solid solutions containing two or more cations in different proportions [5] and these solid solutions are excellent precursors for the synthesis of oxides since the diffusion distances are considerably lower than in the ceramic procedure (Fig. 4.1). The rhombohedral unit cell parameter, a_R, of the carbonate solid solutions varies systematically with the weighted mean cation radius [Fig. 4.2]. Carbonate solid solutions are ideal precursors for the synthesis of monoxide solid solutions of rock-salt

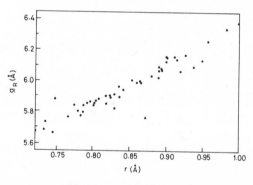

Fig. 4.2 Plot of the rhombohedral lattice parameters, a_R, of a variety of binary and ternary carbonates of calcite structure (e.g. Ca-M, Ca-M-M, Mg-M, M-M′ where M, M′ = Mn, Fe, Co, Cd, etc) against the mean cation radius (Rao et al., *J. Mat. Res.*, 1986)

structure. For example, the carbonates are decomposed in vacuum or in flowing dry nitrogen, to obtain monoxides of the type $Mn_{1-x}M_xO$ (M = Mg, Ca, Co or Cd) of rock-salt structure. Oxide solid solutions of Mg, Ca and Co require temperatures of 770-970K for their formation while those containing cadmium are formed at lower temperatures. The facile formation of oxides of rock-salt structure by the decomposition of carbonates of calcite structure is due to the close (topotactic) relationship between the structures of calcite and rock-salt. The monoxide solids solutions can be used as precursors for preparing spinels and other complex oxides.

Besides monoxide solid solutions, a number of ternary and quarternary oxides of novel structures can be prepared by decomposing carbonate precursors containing the different cations in the required proportion. Thus, one can prepare $Ca_2Fe_2O_5$ and $CaFe_2O_4$ by heating the corresponding carbonate solid solutions in air at 1070 and 1270K respectively for about 1h. $Ca_2Fe_2O_5$ is a defect perovskite with ordered oxide ion vacancies and has the well-known brownmillerite structure (Fig. 4.3) with the Fe^{3+} ions in alternate octahedral (O) and tetrahedral (T) sites. Cobalt oxides of similar compositions, $Ca_2Co_2O_5$ and $Ca_2Co_2O_4$, have been prepared by decomposing the appropriate carbonate precursors around 940K. Unlike in $Ca_2Fe_2O_5$, anion-vacancy ordering in $Ca_2Mn_2O_5$ gives rise to a square-pyramidal coordination (SP) around the transition metal ion (Fig. 4.3). One can also synthesize quarternary oxides, Ca_2FeCoO_5, $Ca_2Fe_{1.6}Mn_{0.4}O_5$, $Ca_3Fe_2MnO_8$ etc. belonging to the $A_nB_nO_{3n-1}$ family, by the carbonate precursor route. In the Ca-Fe-O system, there are several other oxides such as $CaFe_4O_7$, $CaFe_{12}O_{19}$ and $CaFe_2O_4$ $(FeO)_n$ $(n = 1, 2, 3)$ which can, in principle, be synthesized starting from the appropriate carbonate solid solutions and decomposing them in a proper atmosphere.

A good example of a multi-step solid state synthesis achieved starting from carbonate solid solution precursors is provided by the $Ca_2Fe_{2-x}Mn_xO_5$ series of oxides. The structures of both the end members, $Ca_2Fe_2O_5$ and $Ca_2Mn_2O_5$, are derived from that of the perovskite (Fig. 4.3). Solid solutions between the two oxide would be expected to show oxygen vacancy ordered superstructures with Fe^{3+} in octahedral (O) and tetrahedral (T) coordinations and Mn^{3+} in square-pyramidal (SP) coordination, but they cannot be

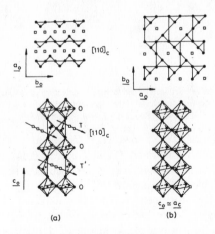

Fig. 4.3 Structures of (a) $Ca_2Fe_2O_5$ (brown millerite) and (b) $Ca_2Mn_2O_5$. Oxygen vacancy ordering in the a–b plane is also shown

prepared by the ceramic method. These solid solutions have indeed been prepared starting from the carbonate solid solutions, $Ca_2Fe_{2-x}Mn_x(CO_3)_4$. The carbonates decompose in air around 1200-1350K to give perovskite-like oxides, $Ca_2Fe_{2-x}Mn_xO_{6-y}$ ($y < 1$). The compositions of the perovskites obtained with $x = 2/3$ and 1 are $Ca_3Fe_2Mn_8$ and $Ca_3Fe_{1.5}Mn_{1.5}O_{8.25}$. X-ray and electron diffraction patterns show that they are members of the $A_nB_nO_{3n-1}$ homologous series with anion-vacancy ordered superstructures with $n = 3$ ($A_3B_3O_{8+x}$). Careful reduction of $Ca_3Fe_2MnO_8$ in dilute hydrogen gives $Ca_3Fe_{4/3}Mn_{2/3}O_5 = Ca_3Fe_2MnO_{7.5}$ (Fig. 4.4). During this step, only Mn^{4+} in the parent oxides is topochemically reduced to Mn^{3+} and Fe^{3+} remains unreduced. The most probable super-structure of $Ca_3Fe_2MnO_{7.5}$ involves SP, O and T polyhedra along the *b*-direction. On

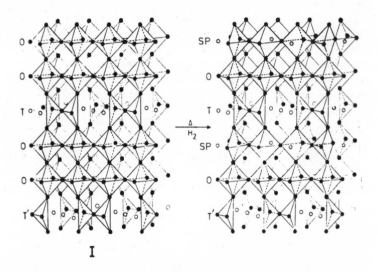

Fig. 4.4 $Ca_3Fe_2MnO_{7.5}$ obtained by the topotactic reduction of $Ca_3Fe_2MnO_8$ (I). The latter is prepared by the decomposition of the precursor carbonate, $Ca_2Fe_{4/3}Mn_{2/3}(CO_3)_4$ (Following Rao et al., *J. Mat. Res.*, 1986)

heating in vacuum at 1140K, however, it transforms to the more stable brownmillerite structure with only O and T coordinations. In Fig. 4.5 we show typical oxides prepared from precursor carbonate solid solutions to illustrate the usefulness of the method.

Ternary and quarternary metal oxides of perovskite and related structures can be prepared by employing hydroxide, nitrate and cyanide solid solutions precursors as well [5]. For example, hydroxide solid solutions of the general formula $Ln_{1-x}M_x(OH)_3$ (where Ln = La or Nd and M = Al, Cr, Fe, Co or Ni)

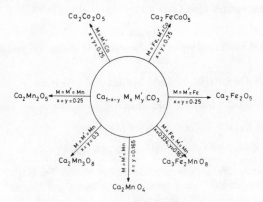

Fig. 4.5   Some of the complex oxides prepared by the decomposition of carbonate precursors

and $La_{1-x-y}M_xM'_y(OH)_3$ (where M = Ni and M' = Co or Cu) crystallizing in the rare earth trihydroxide structure are decomposed at relatively low temperatures (~870K) to yield $LaNiO_3$, $LaNi_{1-x}Co_xO_3$, $LaNi_{1-x}Cu_xO_3$ etc.

Making use of the fact that anhydrous alkaline earth nitrates $A(NO_3)_2$ (A = Ca, Sr, Ba) and Pb $(NO_3)_2$ are isostructural, nitrate solid solutions of the formula $A_{1-x}Pb_x(NO_3)_2$ have been used as precursors for the preparation of ternary oxides such as $BaPbO_3$, Ba_2PbO_4 and Sr_2PbO_4. Quarternary oxides of the type $LaFe_{0.5}Co_{0.5}O_3$ and $La_{0.5}Nd_{0.5}CoO_3$ which cannot be readily prepared by the ceramic method have been obtained by the decomposition of $LaFe_{0.5}Co_{0.5}(CN)_6 \cdot 5H_2O$ and $La_{0.5}Nd_{0.5}Co(CN)_6 \cdot 5H_2O$ respectively [5]. A hyponitrite precursor has been used to prepare superconducting $YBa_3Cu_3O_7$ free from $BaCO_3$ impurity [6].

Chevrel compounds of the general formula $A_xMo_6S_8$ with A = Cu, Pb, La etc (Fig. 1.2) are generally prepared by the ceramic method. A novel precursor compound has been employed [7] to obtain these compounds by a one-step reduction as given by the reaction,

$$2A_x(NH_4)_y Mo_3S_9 + 10 H_2 \longrightarrow A_{2x}Mo_6S_8 + 10H_2S$$
$$+ 2yNH_3 + yH_2$$

Ammonium thiomolybdate, $(NH_4)_2MoS_4$, was reacted with the metal chloride (AX_n) to obtaine the precursor compound. Metal thiolates, thiocarbonates and dithiocarbonates are good precursors for sulfides (e.g. CdS, ZnS). Similar precursor compounds can be thought of for II-VI compounds. Organometallic precursors have been used widely for the synthesis of semiconducting compounds such as GaAs and InP, specially by vapour phase epitaxy (see Table 4.1).

Table 4.1

Typical reactions of organometallic precursors employed in preparing semiconductors[a]

GaAs, $R_3Ga + AsH_3$; GaAsAl, $Me_3Ga + Me_3Al + AsH_3$

GaSb, $Me_3Ga + Me_3$ Sb; InP, $Et_3In + PH_3$

GaInAs, $R_3Ga + R_3Ga + AsH_3$ (or Me_3As)

HgCdTe, $Et_2Te + Me_2Cd + Hg$; ZnSe, $Me_2Zn + Et_2Se$

(a) The precursors are generally alkyls (R = Me/Et). See G.B. Stringfellow, "Organometallic Vapour - phase epitaxy: Theory and Practice", Academic Press, New York, 1987, for details.

(b) Single source compounds are also known: GaAs, Me_2Ga (μ-t-Bu_2As); AsAl, Et_2Al (μ-t-Bu_2As)$_2$, InP, Me_2In (μ-R_2P) and GaSb, $(Gacl_2 \, Sb$t$bu_2)_3$.

Precursor solid solutions or compounds can be used to prepared metal alloys. Thus Mo-W alloys have been prepared [8] by the hydrogen reduction of $(NH_4)_6$ $[Mo_{7-x}W_xO_{24}]$. Metal alloys can be used as precursors to obtain the desired oxides on treatment with oxygen under appropriate conditions. For example, a Eu-Ba-Cu alloy has been oxidized at 1170K to obtain superconducting $EuBa_2Cu_3O_7$ [9].

Organometallic precursors are used in the synthesis of a variety of ceramics, and super-conducting cuprates, being specially necessary for MOCVD. In the case of oxide films, alkoxide or β-diketonate precursors are generally employed. A variety of novel metal alkoxides and related precursor compounds, for the synthesis of oxides as well as for MOCVD of oxides, continue to be prepared. Typical of such new materials are oxoalkoxides prepared by Bradley and others (see Section 8). Films of cuprate superconductors and complex ceramic oxides have thus been prepared by the decomposition of mixtures of metal β-diketonates (e.g. dipivolylmethane) in MOCVD and other techniques. Precursors for BaO, Tl_2O_3 and ZrO_2 have been discussed recently [10a]. Thiolates as sulfide precursors and Cd chalcogenato complexes as precursors of films of II-VI compounds have been described along with precursors of boron and silicon carbides and nitrides [10a]. Special issues of the European Journal of Solid State and Inorganic Chemistry and the Journal of organometallic chemistry [10] have been devoted to organometallic precursors for the synthesis of inorganic materials.

The use of preceramic polymers and other precursors for the synthesis of ceramic materials (especially non-oxide ceramics) has attracted considerable attention in recent years [11]. Verbeek [12] found that thermolysis of CH_3 Si $(NHCH_3)_3$ (prepared by the reaction of CH_3NH_2 with CH_3SiCl_3) around 790K gives a solid carbosilazane resin (soluble in organic solvents) which could be melt-spun at 490K to give fibers. After rendering them infusible by heating in moist air, pyrolysis in N_2 at 1770K gives amorphous ceramic fibers which crystallized on heating to 2070K, to β-SiC with small quantities of α-SiC and β-Si_3N_4. Penn et al [13] have published a detailed study of the preparation and pyrolysis. By using NH_3 instead of CH_3NH_2, a product of the type $(CH_3SiN_{1.5})_n$ was obtained [14] and this was used to obtain polysilazane fibers. A preceramic polymer process for SiC was developed by Verbeek and Winter [15] based on the decomposition of methylchlorosilanes and tetramethylsilane. Yajima and coworkers [16, 17] have done remarkable work on the polymeric precursors of SiC making use of dimethyldichlorosilane as the starting material. Dechlorination by Na results in poly (dimethylsilane), $[Si (CH_3)_2]_n$, which is a white powder; this on heating at around 720K gives a polycarbosilane with a Si-CH_2 backbone which can be melt-spun. On pyrolysis, these fibers give ceramic fibers containing Si, C and O (generally SiC : C : SiO_2 of 1.0 : 0.78 : 0.22). This product is commercially sold as Nicalon. A variation of this fiber containing Si, Ti, C and O has been prepared by heating polycarbosilane with $(n-C_4H_9O)_4$ Ti [17]. A polymer with a C : Si ratio of unity, $[H_2SiCH_2]_n$, has been reported by Wu and Interrante [18].

Silicon nitride has been obtained by the pyrolysis (in a stream of NH_3) of the perhydropolysilazane prepared by the ammonialysis of the H_2SiCl_2-pyridine adduct [19, 20]. The gas stream employed during the pyrolysis of the preceramic polymer plays a crucial role [21]. Pyrolysis of $[B_{10}H_{12}$ diamine$]_n$ polymers in a NH_3 stream gives BN [22]. TiN is similarly obtained by the pyrolysis of an amine precursor [23]. TiN has been prepared from titanazane [23]. Pyrolysis of Nicalon in NH_3 is reported to give Si_3N_4 [24]. Besides single phase ceramics, multiphase ceramics (e.g. composites of SiC and TiC, BiN and Si_3N_4) have been prepared from precursors [25, 26].

There has been considerable work in recent years on many silicon carbide and silicon oxycarbide precursors such as polycarbosilanes, polysiloxanes, aluminium-containing organosilicon polymers and transition metal-containing organ silicon polymers as well as on precursors for Si_3N_4, silicon carbonitride and silicon oxynitride. The main target ceramics in these efforts however remain Si_3N_4 and SiC. The structure and properties of the fibers obtained from the pyrolysis of organosilicon polymers have been reviewed [27]. Precursors for B_4C, BN and boron carbide nitride have also been discussed; disilylproane is a precursor for hydrogenated amorphors SiC [10a]. Titanium carbonitride coatings are formed by the plasma discharge decomposition of titanium dialkylamide.

Laine et al. [28] have described a process where SiO_2 is directly reacted with ethylene glycol and an alkali to produce reactive pentacoordinate silicates which can be used to produce silicate materials. Laine has made stable precurser polymers (> 670 K), some of which are liquid crystalline, by using catechol. Agaskar [29] has prepared organolithic macromolecular materials which are hybrids containing silicate and organic molecules (functionalized spherosilicates) and can be used as precursors for microporous ceramic (Si-C-O) materials.

References

1. C.D. Chandler, C. Roger and M.J.H. Smith, *Chem Rev.* **93** (1993) 1205
2. M.M.A. Sekar and K.C. Patil, *Mater Res. Bull.,* **28** (1993) 485
3. Ch. Laurent, A. Rousset, M. Verelst, K.R. Kannan, A.R. Raju and C.N.R. Rao, *J.Mater. Chem.* **3** (1993) 513
4. L.V. Interrante and A.G. Williams, *Polymer Prepr.* (Am. Chem. Soc. Div. Polym. Chem.) **25** (1984) 13
5. C.N.R. Rao and J. Gopalakrishnan, *Acc. Chem. Res.* **20** (1987) 20
6. H.S. Horowitz, S.J. McLain and A.W. Sleight, *Science* **243** (1989) 66
7. K.S. Nanjundaswamy, N.Y. Vasantacharya, J. Gopalakrishnan, C.N.R. Rao, *Inorg. Chem.* **26** (1987) 4286
8. A.K. Cheetham, *Nature,* **228** (1980) 469
9. K. Matsuzaki, A. Inone, H. Kimura, K. Aoki and T. Masumoto, *Jpn. J. Appl. Phys.,* **26** (1987) L 1310
10. (a) Special issue on precursors for CVD and MOCVD, *Eur. J. Solid State Inorg. Chem.* (H.W. Roesky, ed.) **29** (1992)
    (b) *J. Organometallic Chem.* **449** (1993) Nos 1-2.
11. L.L. Hench and D.R. Ulrich (eds.)., *Science of Ceramic Chemical Processing.* John Wiley, New York, 1986
12. W. Verbeek, U.S. Patent 3853 567 (1974)
13. B.G. Penn, F.E. Ledbetter III, J.M. Clemons and J.G. Daniels, *J. Appl Polym. Sci.* **27** (1982) 3751
14. G. Winter, W. Verbeek and M. Mansmann, U.S. Patent 3892 583 (1975)
15. W. Verbeek and G. Winter, Ger Offen. 2236 708 (1974)
16. S. Yajima, J. Hayashi, M. Omari and K. Okamura, *Nature* **260** (1976) 683; also see S. Yajima, *Am. Ceram. Soc. Bull.,* **62** (1983) 893.
17. S. Yajima, T. Iwai, T. Yamamura, K. Okamura and Y. Hasegawa, *J. Mater Sci.,* **16** (1981) 1349
18. H.J. Wu and L.V. interrante, *Macromolecules,* **25** (1992) 1840; also see C. Whitmarsh and L.V. Interrante, *Organometallics,* **10** (1991) 1336.
19. T. Isoda, H. Kaya, H. Nishii, O. Funayama and T. Suzuki, *J. Inorg. Organomet. Polym.,* **2** (992) 151
20. D. Seyferth, G.H. Wiseman and C. Prudhomme, *J. Am. Ceram. Soc.,* **66** (1984) C-13

21. H.N. Hau, D.A. Lindquist, J.S. Haggerty and D. Seyferth, *Chem. Mater.,* **4** (1992) 705
22. D. Seyferth and W.S. Rees Jr. *Chem. Mater.,* **3** (1991) 1106
23. D. Seyferth and G. Mignani, *J. Mat. Sci.* Lett., **7** (1988) 487
24. K. Okamura, M. Sato and Y. Hasegawa, *Ceram. Int.,* **13** (1987) 55
25. H. Endo. M. Veki and H. Kubo, *J. Mater. Sci.,* 25 (1990) 2503
26. L.V. Interrante and coworkers, *Ceramic Trans.,* **19** (1991) 3, 19 (Adv. Compos Mater)
27. J. Lipowitz. *J. Inorg. Organomet Polymn.* **1** (1991) 277.
28. R.M. Laine et al. *Nature.* *353* (1991) 642; Also R.M. Laine, Abstr. 2nd ANIAC (Asian Chem. Congr.), Malayaisa 1993
29. P.A. Agaskar, *J. Chem. Soc. Chem. Commun.* (1992) 1025

5

Combustion synthesis

Combustion synthesis or the self-propagating high-temperature synthesis is a versatile method for the synthesis of a variety of solids. The method makes use of a highly exothermic reaction between the reactants to produce a flame due to spontaneous combustion (Fig. 5.1) which then yields the desired product or its precursor in finely divided form. Borides, carbides, oxides, chalcogenides and other metal derivatives have been prepared by this method and the topic has been reviewed recently by Merzhanov [1]. In order for combustion to occur, one has to ensure that the initial mixture of reactants is

Fig. 5.1   Combustion reaction during the preparation of a cuprate superconductor (from Rao and Mahesh)

highly dispersed and contains high chemical energy. For example, one may add a fuel and an oxidizer in preparing oxides by the combustion method, to yield the product or its precursor. Thus, one can take a mixture of nitrates (oxidizer) of the desired metals along with a fuel (e.g. hydrazine, glycine or urea) in solution, evaporate the solution to dryness and heat the resulting solid to around 423K to obtain spontaneous combustion, yielding an oxidic product in fine particulate form. Even if the desired product is not formed just after combustion, the fine particulate nature of the product facilitates its formation on further heating.

In order to carry out combustion synthesis, the powdered mixture of reactants (0.1-100 μm particle size) is generally placed in an appropriate gas medium that favors an exothermic reaction on ignition (In the case of oxides, air is generally sufficient). The combustion temperature is anywhere between 1500 and 3500K depending on the reaction. Reaction times are very short since the desired product results soon after the combustion. A gas medium is not always necessary. This is so in the synthesis of borides, silicides and carbides where the elements are quite stable at high temperatures (e.g. $Ti + 2B \longrightarrow TiB_2$). Combustion in a nitrogen atmosphere yields nitrides. Nitrides of various metals have been prepared in this manner. Azides have been used as sources of nitrogen.

The following are typical combustion reactions:

$$MoO_3 + 2\,SiO_2 + 7\,Mg \longrightarrow MoSi_2 + 7\,MgO$$

$$WO_3 + C + 2\,Al \longrightarrow WC + Al_2O_3$$

$$TiO_2 + B_2O_3 + 5\,Mg \longrightarrow TiB_2 + 5\,MgO$$

$$Ta \xrightarrow{\;N_2\;} Ta_2N \xrightarrow[\text{after burning}]{\;N_2\;} TaN$$

MoS_2 and other refractories have been prepared starting from halides [2].

Use of the combustion method in an atmosphere of air or oxygen to prepare complex metal oxides seems obvious [3, 4]. In the last three to four years, a large number of oxides have been prepared by using nitrate mixtures with a fuel such as glycine, urea and teraformalhydrazine. Fine particulate oxide products obtained by this method (Fig. 5.2) may have to be heated further (as in the ceramic method) to yield the desired product (e.g. cuprates). In some cases, the desired oxide is directly obtained. It seems that almost any ternary or quaternary oxide can be prepared by this method. All the superconducting cuprates have been prepared by this method, although the resulting products in fine particulate form, had to be heated at an appropriate

Fig. 5.2 $Y_3Fe_5O_{12}$ powder resulting from the combustion reaction (from K.C. Patil)

high temperature in a desired atmosphere to obtain the final cuprate [4]. In Table 5.1 we list a few of the materials prepared by the combustion method.

Table 5.1

Typical materials prepared by the Combustion Method

Oxides :	$BaTiO_3$, $LiNbO_3$, $PbMoO_4$, $Bi_4Ti_3O_{12}$, $BaFe_{12}O_{19}$, $YBa_2Cu_3O_7$
Carbides :	TiC, Mo_2C, NbC
Borides:	TiB_2, CrB_2, MoB_2, FeB
Silicides:	$MoSi_2$, $TiSi_2$, $ZrSi_2$
Phosphides:	NbP, MnP, TiP
Chalcogenides:	WS_2, MoS_2, $MoSe_2$, TaS_2, $LaTa_3$
Hydrides:	TiH_2, NdH_2

References

1. A.G. Merzhanov in *Chemistry of Advanced Materials* (C.N.R. Rao, Ed.), Blackwell, Oxford, 1992.
2. P.R. Bonneau, R.F. Jarvis Jr. and R.B. Kaner, *Nature, 349* (1991) 510.
3. M.M.A. Sekar and K.C. Patil, *Mater. Res. Bull., 28* (1993) 485.
4. R. Mahesh, V.A. Pavate, Om Parkash and C.N.R. Rao, *Supercond Sci. Tech.,* 5 (1992) 174.

Topochemical reactions

A solid state reaction is said to be topochemically controlled when the reactivity is controlled by the crystal structure rather than by the chemical nature of the constituents. The products obtained in many solid state decompositions are determined by topochemical factors specially when the reaction occurs within the solid without the separation of a new phase [1 - 4]. In topotactic solid state reactions, the atomic arrangement in the reactant crystal remains largely unaffected during the course of the reaction, except for changes in dimension in one or more directions. Orientational relations between the parent and the product phases are generally found. For example, dehydration of β-Ni(OH)$_2$ to NiO and the oxidation of β-Ni(OH)$_2$ to NiOOH are both topochemical reactions; in the former, the orientational relations are (001) ∥ (111) and (110) ∥ (110) while in the latter they are (001) ∥ (001) and (110) ∥ (110). Reduction of NiO to Ni metal also appears to be topochemical [4, 5]. Dehydration of MoO$_3$.2H$_2$O to give MoO$_3$.H$_2$O and the subsequent dehydration to give MoO$_3$ are topochemical [6]. In Fig. 6.1 we show the dehydration of MoO$_3$.2H$_2$O to MoO$_3$.2H$_2$O to MoO$_3$H$_2$O. Dehydration of many other hydrates such as WO$_3$.H$_2$O, VOPO$_4$.2H$_2$O and

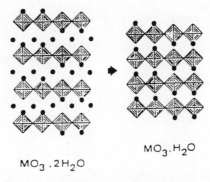

$MO_3 . 2H_2O$

$MO_3 . H_2O$

Fig. 6.1   Dehydration of MoO$_3$.2H$_2$O to MoO$_3$H$_2$O (following Günter, 1972)

HMoO$_2$PO$_4$.H$_2$O occur topochemically. γ-FeOOH topochemically transforms to γ-Fe$_2$O$_3$ on treatment with an organic base. Intercalation and ion exchange reactions are generally topochemical in nature. Reduction of WO$_3$, MoO$_3$ or TiO$_2$ to give lower mixed valent oxides (Magneli phases) is accommodated by the collapse of the structure in specific crystallographic directions. Decomposition of V$_2$O$_5$ to form V$_6$O$_{13}$ is a similar reaction. Reduction of V$_2$O$_5$ by NH$_3$ or a hydrocarbon gives a metastable phase of VO$_2$ in a B′ monoclinic structure similar to that of V$_6$O$_{13}$, which on cooling transforms to the stable monoclinic (B) structure [7]. In Fig. 6.2 we show the reduction of V$_2$O$_5$ dispersed on TiO$_2$ support to give metastable VO$_2$ (B′).

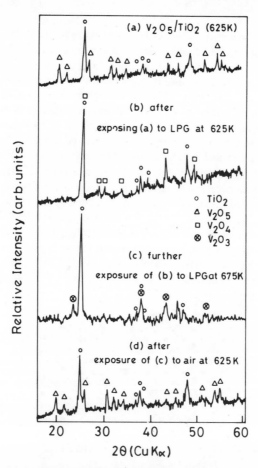

Fig. 6.2 X-ray diffraction patterns of 20 mol% V$_2$O$_5$ dispersed in TiO$_2$ support: (a) at 625K in air; (b) after exposure to liquefied petroleum gas (LPG) at 625K (Notice VO$_2$ (B′) peaks); (c) further exposure of (b) at 675K (Notice V$_2$O$_3$ peaks); (d) after exposure of (c) to air at 625K (The process is fully reversible) (From Raju and Rao 1992)

Many of the modern developments in solid state chemistry owe much to the investigations carried out on MoO_3 and WO_3 (e.g. crystallographic shear planes). WO_3 crystallizes in ReO_3-like structure, but MoO_3 possesses a layered structure (Fig. 6.3). MoO_3 can be stabilized in the WO_3 structure by partly substituting tungsten for molybdenum. $Mo_{1-x}W_xO_3$ solid solutions can be prepared by the ceramic method (by heating MoO_3 and WO_3 in sealed tubes around 870K) or by the thermal decomposition of mixed ammonium metallates. These methods however do not always yield monophasic products owing to the difference in volatilities of MoO_3 and WO_3. The $Mo_{1-x}W_xO_3$ solid solutions are conveniently prepared by the topochemical dehydration of the hydrates [8], the process being very gentle. $MoO_3.H_2O$ and $WO_3.H_2O$ are isostructural and the solid solutions between the two hydrates are prepared readily by adding a solution of MoO_3 and WO_3 in ammonia to hot 6M HNO_3. The hydrates $Mo_{1-x}W_xO_3.H_2O$ crystallize in the same structure as $MoO_3.H_2O$ and $WO_3.H_2O$ with a monoclinic unit cell. The hydrate solid solutions undergo dehydration under mild conditions (around 500K) yielding $Mo_{1-x}W_xO_3$ which crystalize in the ReO_3- related structure of WO_3. The nature of the dehydration of these hydrates has been studied by an *in situ* electron diffraction study where the decomposition occurs due to beam heating [8]. Electron diffraction patterns clearly shown how $WO_3.H_2O$ transforms to WO_3 topochemically with the required orientational relationships. The mixed hydrates, $Mo_{1-x}W_xO_3.H_2O$, undergo dehydration to $Mo_{1-x}W_xO_3$ with similar orientational relations. What is more interesting is that the dehydration of $MoO_3.H_2O$ under electron beam heating gives MoO_3 in the ReO_3 structure, instead of the expected layered structure. The ReO_3 structure of MoO_3 is metastable and can only be produced by the topotactic dehydration under mild conditions. The preparation of ReO_3-like MoO_3 by mild chemical processing is significant. Bulk quantities of MoO_3 in the ReO_3 structure can be prepared by the dehydration of the hydrate [9]. WO_3 1/3 H_2O undergoes topochemical dehydration to yield different phases of WO_3 as shown in Fig. 6.4 [5].

Reduction of ABO_3 perovskites to give $A_2B_2O_5$ and such defect oxides, $ABO_{3-\delta}$, is found to be topochemical (e.g. $CaMnO_3$). The transformation that occurs in such reactions involves the reduction of metal-oxygen octahedra to metal-oxygen square-pyramids (e.g. MnO_5), tetrahedra (e.g. FeO_4) or square-planar units (NiO_4) (see Figs. 4.3 and 4.4). We shall examine the reduction of $LaNiO_3$ and $LaCoO_3$ which crystallize in the rhombohedral perovsksite structure. Occurrence of the $La_nNi_nO_{3n-1}$ homologous series was proposed sometime ago on the basis of a thermogravimetric study of the decomposition of $LaNiO_3$. It was however not known whether a similar series exists in the case of $LaCoO_3$. Controlled reduction of $LaNiO_3$ and $LaCoO_3$ in dilute hydrogen shows the formation of $La_2Ni_2O_5$ and $LaCo_2O_5$ representing the $n = 2$ members of the homologous series LaB_nO_{3n-1}

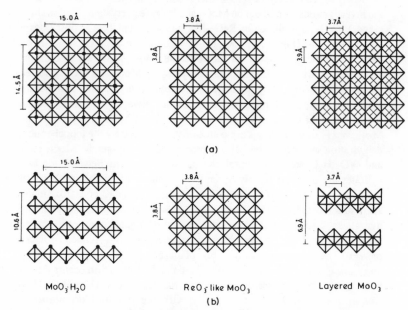

Fig. 6.3   Schematic representation of $MoO_3.H_2O$. MoO_3 in ReO_3-like structure and the layered structure of MoO_3: (a) along [010]; (b) along [001]

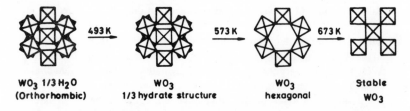

Fig. 6.4   Different WO_3 phases obtained by the dehydration of WO_3 $1/3H_2O$ at different temperatures (From M. Figlarz)

(B = Co or Ni) [10]. $La_2Ni_2O_5$ can only be prepared by the reduction of $LaNiO_3$ at 600K in pure or dilute hydrogen. similarly $La_2Co_2O_5$ can only be prepared by the reduction of $LaCoO_3$ in dilute hydrogen at 670K. Both the oxides can be oxidized back to the parent pervoskites at low temperatures. Neither $La_2Ni_2O_5$ nor $La_2Co_2O_5$ can be prepared by the solid state reaction of La_2O_3 and the transition metal oxide. Reller et al [11] have investigated the anion-deficient $CaMnO_{3-x}$ system obtained by the reduction of $CaMnO_3$.

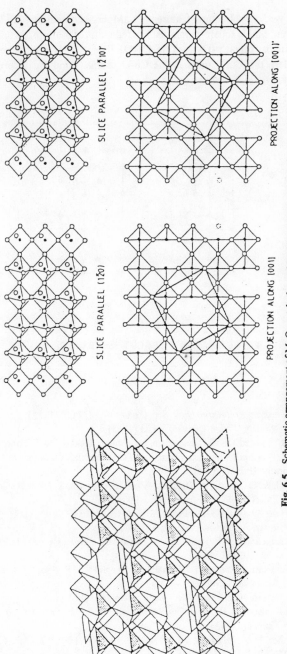

Fig. 6.5 Schematic arrangement of MnO_6 octahedra and MnO_5 square-pyramids in $CaMnO_{2.8}$ (from Reller et al *J. Phys. Chem.* 1983)

In Fig. 6.5 we show structural features of $CaMnO_{2.8}$ containing both MnO_6 octahedra and MnO_5 square-pyramids. The reduction of the high temperature superconductor $YBa_2Cu_3O_7$ to $YBa_2Cu_3O_6$ (Fig. 6.6) is a topochemical process. Here the CuO_4 units in the Cu-O chains along the b direction transform to O-Cu(I)-O sticks.

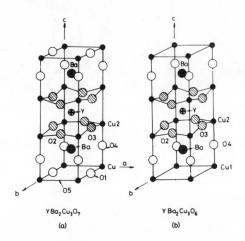

Fig. 6.6   Structures of YBa_2CuO_7 and YBa_2CuO_6

References

1. J.M. Thomas, *Phil. Trans. Roy. Soc. (London), A277* (1974) 251.
2. G.W. Brindley, *Progr. Ceramic Sci.*, 3 (1963).
3. C.N.R. Rao (ed.), Solid State chemistry, Marcel Dekker, New York, 1974;
   C.N.R. Rao and J. Gopalakrishnan, *New Directions in Solid State Chemistry*, Cambridge University Press, 1989.
4. M. Figlarz, B. Gerard, A. Delahaye-Vidal, B. Dumont, F. Harb, A. Coucon and F. Fievet, *Solid State Ionics*, 43. (1990) 143.
5. A. Revolevachi and G. Dhalenne, *Nature*, 316 (1985) 335.
6. J. R. Günter, *J. Solid State Chem.*, 5 (1972) 354.
7. A. R. Raju and C.N.R. Rao, *J. Chem. Soc. Chem. Commun* (1991) 1260; also *Talanta* 39 (1992) 1543.
8. L. Ganapathi, A Ramanan, J. Gopalakrishnan and C.N.R. Rao, *J. Chem. Soc. Chem. Commun.*, (1986) 62.
9. E.M. McCarron, *J. Chem. Soc. Chem. Commun* (1986) 336.
10. K. Vidyasagar, A. Reller, J. Gopalakrishnan and C.N.R. Rao, *J. Chem. Soc. Chem. Commun* (1985) 7.
11. A. Reller, D. A. Jefferson, J. M. Thomas and M. K. Uppal, *J. Phys. Chem.*, 87 (1983) 913.

7

Intercalation chemistry

Intercalation reactions of solids involve the insertion of a guest species (ion or molecule) into a solid host lattice without any major rearrangement of the solid structure.

$$x \text{ (guest)} + \square_x \text{ (host)} \rightleftharpoons (\text{Guest})_x [\text{Host}]$$

where $\square$ stands for a vacant site. Graphite is a well known host which incorporates a variety of guest molecules. Typical intercalation reactions of graphite are the following:

$$\text{Graphite} \xrightarrow[298 \text{ K}]{\text{HF/F}_2} C_x F \quad (x = 3.6 - 4.0)$$

$$\text{Graphite} \xrightarrow[720 \text{ K}]{\text{HF/F}_2} C_x F \quad (x = 0.68 - 1.0)$$

$$\text{Graphite} + x \text{ FeCl}_3 \longrightarrow \text{Graphite } (\text{FeCl}_3)_x$$

$$\text{Graphite} + \text{Br}_2 \longrightarrow C_8 Br$$

$$\text{Graphite} + \text{Conc . H}_2\text{SO}_4 \longrightarrow C_{24}^+ (\text{HSO}_4)^- . 2\text{H}_2\text{SO}_4$$

$$\text{Graphite} \xrightarrow[\text{Vapour} \atop \text{or melt}]{\text{K}} C_8 K$$

$$C_8 K \text{ (bronze)} \xrightarrow[\text{Vacuum}]{} C_{24} K \text{ (steel blue)}$$

$$C_{24} K \longrightarrow C_{36} K \longrightarrow C_{60} K$$

Redox intercalation reactions (e.g. $Li_x TiS_2$ where the lithium metal reduces the TiS_2 layers) can be written as,

$$x \, (guest)^+ + x \, e^- + \square_x \, [host] \rightleftharpoons (Guest)_x \, [Host]$$

A variety of layered structures act as hosts. The general feature of these structures is that the interlayer interactions are weak while the intralayer bonding is strong. Intercalation compounds show interesting phase relations, staging (Fig. 7.1) being an important feature in some of them (e.g. graphite-$FeCl_3$). Higher stages correspond to lower guest concentrations.

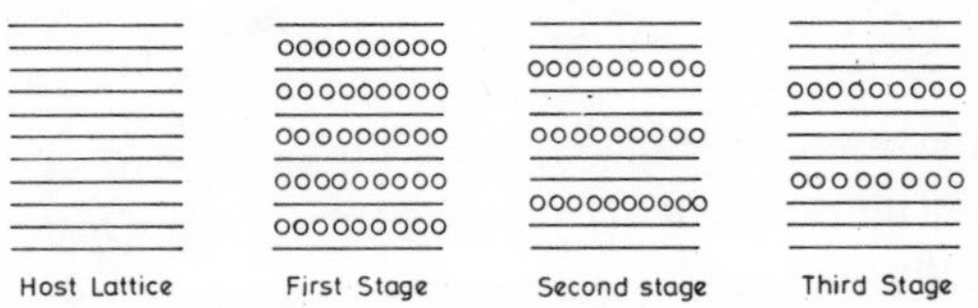

| Host Lattice | First Stage | Second stage | Third Stage |

Fig. 7.1 Staging in intercalation compounds (schematic). Guest molecules are represented by circles in between the layers (shown by lines).

Intercalation chemistry has been reviewed extensively [1-3] and we shall discuss the essential features of these compounds with typical examples.

Alkali metal intercalation involving a redox reaction is readily carried out electrochemically by using the host (MCh_2 dichalcogenide) as the cathode, the alkali metal as the anode and the non-aqueous solution of the alkali metal salt as the electrolyte:

Na/NaI - Propylene carbonate / MCh_2 (Ch = S, Se, Te)

Li/$LiClO_4$ - dioxolane /MCh_2

The reaction is spontaneous if a reverse potential is applied or the cell is short-circuited. Low alkali metal concentrations are obtained by using solutions of salts such as Na or K naphthalide in THF or *n*-butyllithium in hexane.

$$x \, C_4H_9Li + TiS_2 \longrightarrow Li_x \, TiS_2 + \frac{x}{2} \, C_8 H_{18}$$

Alkali metal intercalation in dichalcogenides is also achieved by direct reaction of the elements around 1070K (e.g., A_xMCh_2 where M = V, Nb or Ta) in sealed tubes. Alkali metal intercalation compounds with dichalcogenides form hydrated phases, $A_x(H_2O)_y MCh_2$, just as some of the layered oxides (e.g., $VOPO_4$. $2H_2O$, $MoO_3.2H_2O$). Divalent cations such as Mg^{2+} have been intercalated to TiS_2 using organometallic reagents [5]. NH_3 is intercalated to dichalcogenides by direct reaction (by distilling liquid NH_3

into the dichalcogenide). Intercalation of organic compounds in dichalcogenides is carried out by thermal reaction at temperatures up to 470K. The reaction is generally carried out in neat organic liquids or in a solvent such as benzene or toluene. Amines, amides and pyridine are the types of molecules generally intercalated. Organometallics such as cobaltocene are also intercalated.

Metal phosphorus trisulfides undergo redox intercalation reactions just as the dichalcogenides and also ion exchange reactions giving, $(Guest)^+_x$ $[M_{1-x/2} \square_{x/2} PS_3]$. Metal oxyhalides (e.g. $FeOCl$, WO_2Cl_2) show intercalation reactions similar to dichalcogenides. In addition, they undergo irreversible substitution reactions wherein the halogens of the top layer are substituted by other groups such as $NHCH_3$ and CH_3. Layered metal oxides such as MoO_3, V_2O_5, $MOPO_4$ and $MoAsO_4$ (M = V, Nb, Ta) show reduction reactions similar to the dichalcogenides. Layered oxides of the type AMO_2, $HTiNbO_5$ and $H_2Ti_4O_9$ undergo ion-exchange and oxidative deintercalation reactions. Organic molecules such as amines are intercalated in layered oxides as well. Sheet silicates (e.g. pyrophyllite family, smectites) are also good hosts for organic molecules.

Ready deintercalation of Li from $LiMO_2$ (M = transition metal) enables these materials to be used as cathodes in lithium cells. Delithiation occurs not only by electrochemical methods, but also by reaction with I_2 or Br_2 in solution phase. Lithium insertion in close-packed oxides such as TiO_2, ReO_3, Fe_2O_3 Fe_3O_4 and Mn_3O_4 results in interesting structural changes. Accordingly, 12-coordinated cavities in the ReO_3 framework each becomes two octahedral cavities occupied by lithium. Lithium insertion in Fe_2O_3 changes the anion array from hexagonal to cubic close packing. Jahn-Teller distortion in Mn_3O_4 is suppressed by Li insertion. Lithium-intercalated anatase, $Li_{0.5}TiO_2$, transforms to superconducting $LiTi_2O_4$ at 770K. Delithiation gives rise to oxides in unusual metastable structures (e.g. VO_2 obtained from delithiation of $LiVO_2$). Delithiation of $LiVS_2$ gives VS_2 which cannot otherwise be prepared. In Table 7.1, we list the important hosts and guests in intercalation compounds. In Table 7.2, we list lithium intercalated compounds to show the variety in this system.

Intercalation of sodium and potassium differs from that of lithium. In layered A_xMX_n, lithium is always octahedrally coordinated, while sodium and potassium occupy octahedral or trigonal prismatic sites; octahedral coordination is favoured by large values of x and low formal oxidation states of M. For smaller x and higher oxidation states of M, the coordination of sodium and potassium is trigonal prismatic. Intercalated caesium in MX_n is always trigonal prismatic. Intercalation of sodium and potassium in layered MX_2 oxides and sulfides results in structural transformations involving a change in the sequence of anion layer stacking.

Table 7.1

Examples of hosts and guests in intercalation compounds

Neutral layer hosts	Guests
Graphite	$FeCl_3$, K, Br_2
MCh_2 (M = Ti, Zr, Nb, Ta, etc Ch = S, Se, Te)	Li, Na, NH_3 organic amines, $CoCp_2$
$MPCh_3$ (M = Mg, V, Fe, Zn, etc.)	Li, $CoCp_2$
MoO_3, V_2O_5	H, Alkali metal
$MOPO_4$ and $MOAsO_4$ (M = V, Nb, Ta)	H_2O, Pyridine, Li
$MOCl$ and $MOBr$ (M = V, Fe, etc)	Li, $FeCp_2$
WO_2Cl_2	Zn
Negatively charged layers	
(A) MX_2 (M = Ti, V, Cr, Fe, X = O, S)	A = Group IA (Li, Na.)

Layered silicates and clays
    M $(HPO_4)_2$ (M = Ti, Zr...)
    $K_2Ti_4O_9$
	Organics

Tungsten and molybdenum bronzes, $A_x WO_3$ and $A_x MoO_3$ (A = K, Rb, Cs) are generally prepared by the reaction of the alkali metals with the host oxide; electrochemical methods are also employed for these preparations. Accordingly, $Na_x WO_3$ is prepared by the direct reaction of Na with WO_3 in a sealed tube or by the high temperature reaction (~1270K) of Na_2WO_4 and WO_3 or by electrochemical means. A novel reaction that has been employed to prepare bronzes which are otherwise difficult to obtain involves the reaction of the oxide host with anhydrous alkali iodides [4]:

$$Mo_{1-x}W_xO_3 + y\,(AI) \longrightarrow A_yMo_{1-x}W_xO_3 + \frac{y}{2}I_2$$

Titration of the iodine directly gives the amount of alkali metal intercalated. Atomic hydrogen has been inserted into many binary and ternary oxides. Recently, iodine has been intercalated into the superconducting cuprate, $Bi_2CaSr_2Cu_2O_8$, causing an expansion of the *c*-parameter of the unit cell, without destroying the superconductivity [18]. Note that the oxidation/reduction of $YBa_2Cu_3O_6/YBa_2Cu_3O_7$ is an intercalation reaction. Chevrel compounds, $A_xMo_6Ch_8$ (Ch = S, Se or Te), may be considered to be intercalation compounds. Thus Mo_6S_8 is prepared by acid-leaching Cu from $Cu_xMo_6S_8$. Alkali fullerides of the type A_3C_{60} (A = alkali metal) can also be considered to be intercalation compounds. In fact, A_3C_{60} can be made by the intercalation of A into C_{60} at low temperatures from liquid ammonia solution [18a].

Table 7.2

Intercalation compounds of Lithium

Host	Description	Reference
TiS_2	Li_xTiS_2 $(0 < x \leq 1)$	[6]
VS_2	Li_xVS_2 $(0 < x \leq 1)$ phases obtained by deintercalation of lithium from $LiVS_2$ using I_2/CH_3CN. Three different phase regions: $0.25 \leq x \leq 0.33$; $0.48 \leq x \leq 0.62$ and $0.85 \leq x \leq 1$ apart from VS_2	[7]
NbS_2 (3R)	$Li_{0.5}NbS_2$ and $Li_{0.70}NbS_2$	[6]
MoS_3	Li_xMoS_3 $(0 < x \leq 4)$	[8]
MO_2	Li_xMO_2 $(x \geq 1)$ (M = Mo, Ru, Os or Ir) MO_2 of rutile structure	[9]
TiO_2 (anatase)	Li_xTiO_2 $(0 < x \leq 0.7)$. $Li_{0.5}TiO_2$ transforms irreversibly to $LiTi_2O_4$ spinel at 770 K	[10]
CoO_2	Li_xCoO_2 $(0 < x < 1)$ phase obtained by electrochemical delithiation of $LiCoO_2$	[11]
VO_2	(a) Li_xVO_2 $(0 < x < 1)$ phase obtained by delithiation of $LiVO_2$ using $Br_2/CHCl_3$	[12]
	(b) Li_xVO_2 $(0 < x < 2/3)$; lithiation using n-butyl lithium	[13]
Fe_2O_3	$Li_xFe_3O_4$ $(0 < x < 2)$; anion array transforms from hcp to ccp on lithiation	[14]
Fe_3O_4	$Li_xFe_3O_4$ $(0 < x < 2)$ Fe_2O_4 subarray of the spinel structure remains intact	[14]
Mn_3O_4	$Li_xMn_3O_4$ $(0 < x < 1.2)$; lithium insertion suppresses tetragonal distortion of Mn_3O_4	[14]
MoO_3	Li_xMoO_3 $[0 < x < 1.55)$	[15]
V_2O_5	$Li_xV_2O_5$ $(0 < x < 1.1)$; intercalation of lithium by using LiI	[16]
ReO_3	$Li_x ReO_3$ $(0 < x < 2)$; three phases $0 < x \leq 0.35$, $x = 1$ and $1.8 \leq x \leq 2$	[17]

Deintercalation provides a novel means of obtaining metastable solids in unusual structures (e.g. $B-VO_2$ and VS_2 from $LiVO_2$ and $LiVS_2$). Gopalakrishnan et al [19] have prepared $V_2(PO_4)_3$ of NASICON structure by the oxidative deintercalation of Na from $Na_3V_2(PO_4)_3$ using a halogen in $CHCl_3$ solution.

Since $\alpha-Zr(HPO_4)_2.H_2O$ ($\alpha-ZrP$) was first characterized by Clearfield and Stynes, a number of layered compounds of the general formula M(IV)

$(O_3PR)_x (O_3PR')_{2-x}$ nS where $R \neq R' = OH$, H, CH_3, C_6H_5 etc. have been discovered. These compounds provide great possibilities for interlayer chemistry, including engineering of systems where guest molecules bind depending of their shape and chemical properties [19a]. Compounds of the type Zr (O_3PRPO_3) where adjacent inorganic layers are bound by organic radicals have been prepared [19b]. Compounds of the type $ZrPO_4.RPO_2R'$ have been prepared by topotactic reactions, as also pillared $ZrPO_4.R'$ $O_2P-R-PO_2$ R' [19c]. Potential applications of this class of materials are indeed very large.

Stable colloidal dispersions of various classes of compounds with layered and chain structures can be prepared by appropriately manipulating interlayer or interchain interactions. The dispersion phenomenon is known in smectite clay minerals which readily exfoliate in water to form sols or gels depending on the concentration of the colloidal particles. Several phases with layer charges comparable to the smectite clays are known to exfoliate spontaneously in high dielectric constant solvents. Typical examples are Na_xMS_2 and M_xMPS_3 [20, 21]. Synthesis of single MoS_2 layers by the reaction of the lithium intercalation compound with water (exfoliation) is another interesting example [22]. In Fig. 7.2 we show a schematic representation of a single layer of MoS_2 and such layered chalcogenides

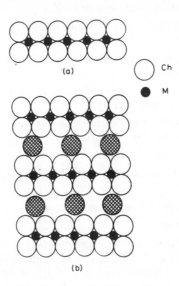

Fig. 7.2   Schematic representation of (a) a single layer of MCh_2, M = Nb, Mo or W and Ch = S, Se) obtained by exfoliation. Black circles are M atoms and open circles are Ch atoms. (b) organic inclusion compound where the hatched circular shapes are organic molecules (Following Diwaigalpitiya et al *Science*, 1989).

obtained by exfoliation and of an inclusion compound obtained by intercalating an organic molecule between the chalcogenide layers. For compounds with higher layer charges, some modification of the interlayer structure becomes necessary to promote the reaction. For this purpose, dispersions have been prepared by ion exchange of small cations such as lithium. The high solvation energy of the lithium cation balances the solid lattice energy, thereby permitting chain or layer separation to occur. Dispersions of $[Mo_3Se_3]^{2-}$ and $[FeS_2]$ chains have thus been prepared [23]. It the interlayer interactions are essentially hydrogen-bonding in nature, partial intercalation of a primary organic amine is useful. For example, partial intercalation of n-propylamine into $Zr(HPO_4)_2.2H_2O$ promotes spontaneous exfoliation by reducing hydrogen bonding between the layers. A stable phase, is observed when complete exchange occurs and strong van der Waals forces between a close packed bilayer of organic groups replace the hydrogen bonds as the dominant interlayer interaction [24].

Other approaches are employed to promote exfoliation in higher layer charges. For example, in perovskite related layer phases, $HCa_2Na_{n-3}Nb_nO_{3n+1}$, a surfactant molecule with an amine head group was first intercalated by protonation. The surfactant generally has a hydrophilic polyether tail which enhances the intercalation of solvent molecules. Stable dispersions in water and other polar solvents have thus been obtained [25]. Clearfield finds that $Zr[O_3PC_6H_4SO_3H_2)$ exfoliates spotaneously in aqueous media and sequesters large cations.

Dispersions can also be flocculated by the addition of electrolytes, giving rise to systems containing large cations along with layers (or chains). Such systems cannot be prepared otherwise by direct intercalation. Intercalation compounds of TaS_2 with complex aluminium oxycations and an iron sulfur cluster have been prepared in water-N-methyl-formamide solution [26].

References

1. M.S. Whittingham and A.J. Jacobson (eds), *Intercalation Chemistry*, Academic Press, New York, 1982.
2. C.N.R. Rao and J. Gopalakrishnan, *New Directions in Solid State Chemistry*, Cambridge University Press, 1989.
3. A.J. Jacobson in *Solid State Chemistry-Compounds* (A.K. Cheetham and P. Day, eds), Clarendon Press, Oxford, 1992.
4. A.K. Ganguli, J. Gopalakrishnan and C.N.R. Rao, *J. Solid State Chem.* 74 (1988) 228.
5. P. Lightfoot, F. Krok, J.L. Nowinski and P.G. Bruce, *J.Mater. Chem.* 2 (1992) 139.
6. M.S. Whittingham, *Prog. Solid State Chem.* 12 (1978) 41.
7. D.W. Murphy, C. Cros, F.J. Di Salvo and J.V. Waszezak, *Inorg. Chem.* 16 (1977) 3027.

8. A.J. Jacobson, R.R. Chianelli, S.M. Rich and M.S. Whittingham *Mater. Res. Bull* **14** (1979) 1437.
9. D.W. Murphy, F.J. DiSalvo, J.N. Carides and J.V. Waszezak, *Mater. Res. Bull.* **13** (1978) 1395.
10. D.W. Murphy, M. Greenblatt, S.M. Zahurak, R.J. Cava, J.V. Waszezak and R.S. Hutton, *Rev. Chim. Miner* **19** (1982) 441.
11. K. Mizushima, P.C. Jones, P.J. Wiseman and J.B. Goodenough, *Mater. Res. Bull.* **15** (1980) 783.
12. K. Vidyasagar and J. Gopalakrishnan, *J.Solid State Chem.* **42** (1982) 217.
13. D.W. Murphy and P.A. Christian, *Science*, **205** (1979) 651.
14. M.M. Thackeray, W.I.F. David and J.B. Goodeough, *Mater. Res. Bull.* **17** (1982) 785; **18** (1983) 461.
15. P.G. Dickens and M.F. Pye in ref. [1].
16. P.G. Dickens, S.J. French, A.T. Hight and M.F. Pye, *Mater. Res. Bull.* **14** (1979) 1295.
17. R.J. Cava, A. Santoro, D.W. Murphy, S. Zahurak and R.S. Roth, *J. Solid State Chem* **42** (182) 251.
18. X.D. Xiang, S.McKernan and W.A. Vareka, *Nature* **348** (1990) 145.
18a. R.C. Haddon, *Acc. Chem. Res.* **25** (1992) 127; B.L. Ramakrishna, Z. Iqbal, E.W. Ong. D. Yang, S.N. Murthy, P. Askeljer, F. Korenivski, K.V. Rao, K. Sinha, J. Menedez, J.S. Kim and R.F. Marzke, To be published.
19. J. Gopalakrishnan and M. Kasturi Rangan, *Chem. Mater,* **4** 745 (1992).
19a. G.H. Hong and T. Mallouk, *Acc. Chem. Res.,* **25** (1992) 420.
19b. G. Alberti, in *Proc. Intnl. Symp. on Soft Chemistry Routes to New Materials,* Nantes, 1993 (Trans Tech Publications).
19c. G. Alberti, M. Casciola and R.K. Biswas, *Inorg. Chem. Acta,* **201** (1992) 207.
20. A. Lerf and R. Schöllhorn, *Inorg. Chem.,* **16** (1977) 2950.
21. R. Clement, O. Garnier and J. Jegoudez, *Inorg. Chem.,* **25** (1986) 404.
22. W.M.R. Diwaigalpitiya, R.F. Frindt and S.R. Morrison, *Science,* **246** (1989) 369.
23. J.M. Tarascon, F.J. DiSalvo, C.H. Chen, P.J. Carroll, M. Walsh, and L. Rupp. *J. Solid State Chem.,* **58** (1985) 290.
24. G. Alberti, M. Casciola and U. Costantino, *J. Colloid Interfacial Sci.* **107** (1985) 256.
25. M.M.J. Treacy, S.B. Rice, A.J. Jacobson and J.T. Lewandowski, *Chem. Mater,* **2** (1990) 279.
26. L.F. Nazar and A.J. Jacobson, *J. Chem. Soc. Chem. Commun.,* (1986) 570.

8

Sol-gel synthesis

The sol-gel method has provided a very important means of preparing inorganic oxides. It is a wet chemical method and a multistep process involving both chemical and physical processes such as hydrolysis, polymerization, drying and densification. The name "sol-gel" is given to the process because of the distinctive viscosity increase that occurs at a particular point in the sequence of steps. A sudden increase in viscosity is the common feature in sol-gel processing, indicating the onset of gel formation. In the sol-gel process, synthesis of inorganic oxides is achieved from inorganic or organometallic precursors (generally metal alkoxides). Most of the sol-gel literature deals with synthesis from alkoxides. Ethyl orthosilicate, $Si(OEt)_4$, titanium tetra-iso-propoxide are typical alkoxides used in sol-gel synthesis.

The important features of the sol-gel method are: better homogeneity compared to the traditional ceramic method, high purity, lower processing temperature, more uniform phase distribution in multicomponent systems, better size and morphological control, the possibility of preparing new crystalline and non-crystalline materials and lastly easy preparation of thin films and coatings. The sol-gel method is widely used in ceramic technology and the subject has been widely reviewed [1-3].

The six important steps in sol-gel synthesis are the following:

Hydrolysis: The process of hydrolysis may start with a mixture of a metal alkoxide and water in a solvent (usually alcohol) at the ambient or a slightly elevated temperature. Acid or base catalsts are added to speed up the reaction.

Polymerization: This step involves condensation of adjacent molecules wherein H_2O and alcohol are eliminated and metal oxide linkages are formed. Polymeric networks grow to colloidal dimensions in the liquid (*sol*) state.

Gelation: In this step, the polymeric networks link up to form a three-dimensional network throughout the liquid. The system becomes somewhat rigid, characteristic of a gel, on removing the solvent from the sol. Solvent as well as water and alcohol molecules, however, remain inside the pores of the gel. Aggregation of smaller polymeric units to the main network progressively continues on aging the gel.

Drying: Here, water and alcohol are removed at moderate temperatures (< 470K), leaving a hydroxylated metal oxide with resi- dual organic content. If the objective is to prepare a high surface area *aerogel* powder of low bulk density, the solvent is removed supercritically.

Dehydration: This steps is carried out between 670 and 1070 K to drive off the organic residues and chemically bound water, yielding a glassy metal oxide with up to 20–30% microporosity.

Densification: Temperatures in excess of 1270 K are used to form the dense oxide product.

The above-mentioned steps in the sol-gel method may or may not be strictly followed in practice. Thus, man complex metal oxides are prepared by a modified sol-gel route without actually preparing metal alkoxides. For example, a transition metal salt (e.g., metal nitrate) solution is converted into a gel by the addition of an appropriate organic reagent (e.g. 2-ethyl-1 hexanol). Alumina gels have been prepared by ageing sols obtained by the hydrolysis of Al s-butoxide followed by hydrolysis in hot H_2O and peptization with HNO_3 [4]. In the synthesis of oxides containing Ti, Zr and such metals, the metal halide ($TiCl_4$, $ZrCl_4$) is taken with ethyl orthosilicate and an organic base (imidazole, pyrolidine, pyrrole etc.). In the case of cuprate superconductors, an equimolar proportion of citric acid is added to a solution of metal nitrates, followed by ethylenediamine until the solution attains a pH of 6-6.5. The blue sol is concentrated to obtain the gel. The *xerogel*, obtained by heating at ~420K, is then decomposed at an appropriate temperature to get the cuprate.

The sol-gel technique has been used to prepare sub-micrometer metal oxide powders [5] with a narrow particle size distribution and unique particle shapes (e.g. Al_2O_3, TiO_2, ZrO_2, Fe_2O_3). Uniform SiO_2 spheres have been grown from aqueous solutions of colloidal SiO_2 [6]. Metal-ceramic composites (e.g. Ni–Al_2O_3, Pt–ZrO_2) can also be prepared in this manner [7]. Organic-inorganic composites have been prepared by the sol-gel route. By employing several variants of the basic sol-gel technique, a number of multicomponent oxide system have been prepared. Typical of them, are, SiO_2–B_2O_3, SiO_2–TiO_2, SiO_2–ZrO_2, SiO_2–Al_2O_3, and ThO_2–UO_2. A variety of ternary and still more complex oxides such as $PbTiO_3$, $PbTi_{1-x}$ Zr_xO_3 and NASICON have been prepared by this techniques [1-3, 8]. Different types of cuprate superconductors have been prepared by the sol-gel method; these include $YBa_2Cu_3O_7$, $YBa_2Cu_4O_8$, $Bi_2CaSr_2Cu_2O_8$ and $PbSr_2Ca_{1-x}Y_xCu_3O_8$ [9].

Efforts to prepare new alkoxides continue and many complex metal alkoxides have been prepared. For example, several metal oxoalkoxide clusters have been prepared by Bradley et al [10]. Aluminium oxoalkoxides such as $Al_4(\mu_4 - O)(\mu_2 - OBu)_5 (OBu_6{}^i) H^2$ and $Al_{10} (OEt)_{22}O_4$ have been reported in the literature [11].

References

1. L.L. Hench and D.R. Ulrich (eds), *Science of Ceramic Chemical Processing*, John Wiley, New York, 1986.
2. J. Livage, M. Henry and C. Sonchez, *Progress in Solid State Chem.*, 77 (1992) 153.
3. D.R. Uhlmann, B.J. Zelinski and G.E. Wnek, *Better Ceramics through Chemistry* (C.J. Brinker, D.E. Clark and D.R. Ulrich, (eds), *Better Ceramics through Chemistry*, IV MRS Symposium 180, (1990) and the earlier volume.
4. B.E. Yoldas, *J. Mater. Res.*, 10 (1975) 1856.
5. E. Matijevic in *Ultrastructure Processing of Ceramics, Glassess and Composites* (L.L. Hench and D. Ulrich (eds.), John Wiley, New York, 1984.
6. R.K. Iler, *The Chemistry of Silica*, John Wiley, New York, 1979.
7. M. Verelst, K.R. Kannan, G.N. Subbanna, C.N.R. Rao, Ch. Laurent and A. Rousset, *J. Mater. Res.* 7 (1992) 3072 and the references listed therin.
8. J. Alamo and R. Roy, *J. Solid State Chem.* 51 (1984) 270.
9. C.N.R. Rao, R. Nagarajan and R. Vijayaraghavan, *Supercond. Sci. Tech.* 6 (1993) 1.
10. D.C. Bradley, H. Chudzynska, D.M. Frigo, M.E. Hammond, M.B. Hursthouse and M.A. Mazid, *Polyhedron*, 9 (1990) 719.
11. R.A. Sinclair, W.B. Gleason, R.A. Newmark, J.R. Hill, S. Hunt, P. Lyon and J. Stevens, in *Chemical Processing of Advanced Materials* (L.L. Hench and J.K. West, eds), John Wiley, New York, 1992, p. 207.

9

Ion Exchange Method

Ion exchange in fast-ion conductors such as β-alumina is well known. It can be carried out in aqueous as well as molten salt media conditions. Accordingly, β-alumina has been exchanged with Li^+, K^+, Ag^+, Cu^+, H_3O^+, NH_4^+ and other monovalent and divalent cations, giving rise to different β-aluminas [1]. This is generally done by immersing β-alumina in a suitable molten salt (around 570 K). Divalent Ca^{2+} is known to replace two Na^+ ions. Ion exchange in inorganic solids is a general phenomenon, not restricted to fast ion conductors alone. For example, $Ag_2Si_2O_5$ with a sheet silicate structure is prepared by immersing $Na_2Si_2O_5$ in molten $AgNO_3$. Kinetic and thermodynamic aspects of ion exchange in inorganic solids were examined sometime ago by England et al [2]. Their results reveal that ion exchange is a phenomenon that occurs even when the diffusion coefficients are as small as $\sim 10^{-12}$ $cm^2 s^{-1}$, at temperatures far below the sintering temperatures of solids. Ion exchange occurs at a considerable rate in stoichiometric solids as well. Mobile ion vacancies introduced by nonstoichiometry or doping seem to be unnecessary for exchange to occur. Since the exchange occurs topochemically, it enables the preparation of metastable phases that are inaccessible by high-temperature reactions.

England et al [2] have shown that a variety of metal oxides possessing layered, tunnel or close-packed structures can be ion-exchanged in aqueous solutions or molten salt media to produce new phases. Typical examples are:

$$\alpha\text{-}NaCrO_2 \xrightarrow[570K; 24h]{LiNO_3} \alpha\text{-}LiCrO_2$$

$$KAlO_2 \xrightarrow{AgNO_3\,(l)} \beta\text{-}AgAlO_2$$

$$\alpha\text{-}LiFeO_2 \xrightarrow{CuCl\,(l)} CuFeO_2$$

The structure of the framework is largely retained during ion exchange except for minor changes to accommodate the structural preferences of the incoming ion. Thus, when α-$LiFeO_2$ is converted to $CuFeO_2$ by exchange with molten $CuCl$, the structure changes from that of α-$NaCrO_2$ to that of delafossite to provide a linear anion coordination for Cu^+. Delafossites, $A^IB^{III}O_2$ (A = Ag, Cu, B = Fe, Ni etc.) are all made by ion-exchange reactions. Similarly, when $KAlO_2$ is converted to β-$AgAlO_2$ by ion exchange, there is a change in structure from cristobalite-type to ordered wurtzite-type. The change probably occurs to provide a tetrahederal coordination for Ag^+.

An interesting ion exchange reaction is the conversion of $LiNbO_3$ and $LiTaO_3$ to $HNbO_3$ and $HTaO_3$ respectively, by treatment with hot aqueous acid [3]. The exchange of Li^+ by protons is accompanied by a topotactic transformation of the rhombohedral $LiNbO_3$ structure to the cubic perovskite structure of $HNbO_3$. The mechanism suggested for the transformation is the reverse of the transformation of cubic ReO_3 to rhombohedral $LiReO_3$ and Li_2ReO_3 [4], involving a twisting of the octahedra along the [111] cubic direction so as to convert the 12-coordinated perovskite tunnel sites to two 6-coordinated sites in the rhombohedral structure. An interesting structural change accompanying ion exchange is found in $Na_{0.7}CoO_2$ [5] where the anion sequence is ABBAA; cobalt ions occur in alternate interlayer octahedral sites and sodium ions in trigonal prismatic coordination in between the CoO_2 units. When this material is ion-exchanged with $LiCl$, a metastable form of $LiCoO_2$ with the layer sequence ABCBA is obtained. The phase transforms irreversibly to the stable $LiCoO_2$ (ABCABC) around 520K.

A variety of inorganic solids have been exchanged with protons to give new phases, some of which exhibit high protonic conduction, typical of them being $HTaWO_6.H_2O$, $HMO_3.xH_2O$ (M = Sb, Nb, Ta), pyrochlores and $HTiNbO_5$ [6]. Ion exchange has also been reported in metal sulphides. For example, $KFeS_2$ undergoes topochemical exchange of potassium in aqueous solutions with alkaline earth metal cations to give new phases in which the $[FeS_{4/2}]$ tetrahedral chain is preserved [7].

References

1. B.C. Tofield in *Intercalation Chemistry* (M.S. Wittingham and A.J. Jacobson, eds), Academic Press, New York, 1982.
2. W.A. England, J.B. Goodenough and P.J. Wiseman, *J. Solid State Chem.*, **49** (1983) 289.
3. C.E. Rice and J.L. Jackal, *J. Solid State Chem.*, **41** (1982) 308.
4. R.J. Cava, A. Santoro, D.W. Murphy, S. Zahurak and R.S. Roth, *J. Solid State Chem.* **42** (1982) 251.

5. C. Delmas, J.J. Braconnier and P. Hagenmuller, *Mater. Res. Bull.* **17** (1982) 117.

6. C.N.R. Rao and J. Gopalakrishnan, *New Directions in Solid State Chemistry*, Cambridge University Press, 1989.

7. H. Boller, *Monatsh Chem.* **109** (1978) 975.

10

Use of Alkali Media

Strong alkaline media, either in the form of solid fluxes or molten (or aqueous) solutions, enable the synthesis of novel oxides. The alkali flux stabilizes higher oxidation states of metals by providing an oxidizing atmosphere. Alkali carbonate fluxes have been traditionally used to prepare transition metal oxides such as $LaNiO_3$ with Ni in the 3^+ state. A good example of an oxide synthesized in a strongly alkaline medium is the pyrochlore, $Pb_2 (Ru_{2-x}Pb_x)O_{7-y}$ where Pb is in the 4^+ state [1]. This oxide is a bifunctional electrocatalyst. The procedure for preparation involved bubbling oxygen through a solution of Pb and Ru salts in strong KOH at 320 K. The so-called alkaline hypochlorite method is used in many instances. For example, $La_4Ni_3O_{10}$ was prepared by bubbling Cl_2 gas through a NaOH solution of lanthanum and nickel nitrates of appropriate stoichiometry [2].

Superconducting $La_2CuO_{4+\delta}$ has been prepared by reacting a mixture of La_2O_3 and CuO in molten KOH-NaOH around 520K [3]. It is possible that alkali metal ions are incorporated in the oxide during the reaction. Alkaline hypobromite oxidation also yields superconducting $La_2CuO_{4+\delta}$ [4]. $LnBa_2Cu_3O_7$ (Ln = Y or Er) has been prepared in a fused NaOH–KOH flux [3, 5]. $YBa_2Cu_4O_8$ has been prepared by using a Na_2CO_3–K_2CO_3 flux in a flowing oxygen atmosphere [6]. KOH melt has been used to prepare superconducting $Ba_{1-x}K_xBiO_3$ [7]. $BaCuO_{2.5}$ in a K_2NiF_4–like structure has been prepared using a $NaOH-Na_2O_2$ molten flux [8].

References

1. H.S. Horowitz, J.M. Longo and J.T. Lewandowski, *Mater Res. Bull.* **16** (1981) 489.
2. R.A. Mohan Ram, L. Ganapathi, P. Ganguly and C.N.R. Rao, *J. Solid State Chem.* **63** (1986) 139.
3. W.K. Ham G.F. Holland and A.M. Stacy, *J. Am. Chem. Soc.* **110** (1988) 5214; L.N. Marquez, S.W. Keller and A.M. Stacy, *Chem. Mater.* **5** (1993) 761.
4. P. Rudolf and R. Schollhorn, *J. Chem. Soc. Chem. commun.* (1992) 1158.

5. N. Coppa, A. Kebede, J.W. Schwegler, I. Perez, R.E. Salomon, G.H. Myer and J.E. Crow *J. Mater. Res.* **5** (1990) 2755.

6. R.J. Cava, J.J. Krajewski and W.F. Peck, *Nature*, **338** (1989) 328.

7. L.F. Schneemeyer, J.K. Thomas and T. Siegrist, *Nature*, **355** (1988) 421.

8. Unpublished results from the author's laboratory.

11

Electrochemical Methods

Electrochemical methods have been employed to advantage for the synthesis of many solid materials [1-4]. Typical of the materials prepared in this manner are metal borides, carbides, silicides, oxides and sulfides as can be seen from the listing in Table 11.1. Vanadate spinels of the formula MV_2O_4 as well as tungsten bronzes, A_xWO_3, have been prepared by the electrochemical route. Tungsten bronzes are obtained at the cathode when current is passed through two inert electrodes immersed in a molten solution of the alkali metal tungstate, A_2WO_4 and WO_3; oxygen is liberated at the anode [5]. Blue Mo bronzes have been prepared by fused salt electrolysis [6]. Oxides containing metals in high oxidation states are conveniently prepared electrochemically (e.g. $La_{1-x}Sr_xFeO_3$). In Fig. 11.1, we have shown a typical electrode system described by Pouchard and coworkers [7].

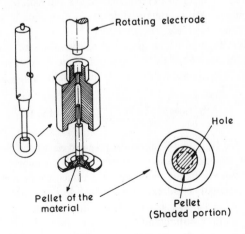

Fig. 11.1 Electrode system employed for electrochemical oxidation (Following Wattiaux et al 1985)

Superconducting $La_2CuO_{4+\delta}$ has been recently prepared by electrochemical oxidation [8]. Ferromagnetic, cubic $LaMnO_3$ with ~45% Mn^{4+} has been prepared electrochemically [9].

Table 11.1

Typical electrochemical preparations[a]

Constituents of melt	Product	T(K)
Na_2WO_4, WO_3	Na_xWO_3	
Na_2MoO_3, MoO_3	MoO_2	945
$CaTiO_3$, $CaCl_2$	$CaTi_2O_4$	1120
NaOH, Ni electrodes	$NaNiO_2$	
$Na_2B_4O_7$, NaF, V_2O_5, Fe_2O_3	FeV_2O_4	1120
$Na_2B_4O_7$, NaF, WO_3, Na_2SO_4	WS_2	1070
$NaPO_3$, Fe_2O_3, NaF	FeP	1195
Na_3CrO_4, Na_2SiF_6	Cr_3Si	
$Na_2Ge_2O_5$, NaF, NiO	Ni_2Ge	
$Li_2B_4O_7$, LiF, Ta_2O_5	TaB_2	1220

(a) from ref [3]

Monosulfides of U, Gd, Th and other metals are obtained from a solution of the normal valent metal sulfide and chloride in a NaCl/KCl eutectic. LaB_6 is prepared by taking La_2O_3 and B_2O_3 in a $LiBO_2$/LiF melt and by using gold electrodes. Crystalline transition metal phosphides are prepared from solutions of oxides with alkali metal phosphates and halides.

As mentioned earlier, intercalation of alkali metals in host solids is readily accomplished electrochemically. It is easy to see how both intercalation (reduction of the host) and deintercalation (oxidation of the host) are processes suited for this method. Thus, lithium intercalation is carried out by using lithium anode and a lithium salt in a non-aqueous solvent.

$$MS_2 \text{ (s)} + xLi^+ + xe^- \rightleftarrows Li_xMS_2 \text{ (s)}$$

Superconducting $Ba_{1-x}K_x BiO_3$ has been prepared electrochemically [10].

Although the electrochemical method has been known for long, the processes involved in the synthesis of various solids are not entirely understood. Generally one uses solvents whose decomposition potentials are high (e.g. alkali metal phosphates, borates, fluorides, etc.). Changes in melt composition could cause limitation in certain instances. There is considerable scope to investigate the chemistry and applications of electrochemical methods of synthesis of solids.

References

1. C.N.R. Rao and J. Gopalakrishnan, *New Directions in Solid State Chemistry*, Cambridge University Press, 1989.
2. J.D. Corbett in *Solid State Chemistry-Techniques,* (A.K. Cheetham and P. Day, eds), Clarendon Press, Oxford, 1987.
3. A. Wold and D. Bellavance in *Preparative Methods in Solid Strate Chemistry* (D. Hagenmuller, ed), Academic Press, New York, 1972.
4. R.S. Feigelson, *Adv. Chem. Sci.* **186** (1980) 243.
5. M.S. Whittingham and R.A. Huggins in *Solid State Chemistry* (R.S. Roth and S.J. Schneider Jr. eds), National Bureau of Standards, 1972.
6. E. Banks and A. Wold in *Solid State Chemistry* (C.N.R. Rao, ed.), Marcel Dekker, New York, 1974.
7. A. Wattiaux, J.C. Grenier, M. Pouchard and P. Hagenmuller, *Rev. Chim. Minerale* **22** (1985) 1.
8. J.C. Grenier, A. Wattiaux, N. Lagueyte, J.C. Park, E. Marquestaut, J. Etourneau and M. Pouchard, *Physica C,* **173** (1991) 139.
9. Unpublished results from the author's laboratory.
10. J.M. Rosamilia, S.H. Glarum, R.J. Cava, B. Batlogg and B. Miller, *Physica C* **182** (1991) 285.

12

Nebulized Spray Pyrolysis

Pyrolysis of sprays is a well-known method for depositing films. Thus, one can obtain films of oxidic materials such as CoO, ZnO and $YBa_2Cu_3O_7$ by the spray pyrolysis of solutions containing salts (e.g. nitrates) of the cations. A novel improvement in this technique is the so-called *pyrosol process* or nebulized spray pyrolysis involving the transport and subsequent pyrolysis of a spray generated by an ultrasonic atomizer as demonstrated by Joubert and coworkers [1]. Wold and coworkers as well as Rao and coworkers have employed this method to prepare films of a variety of oxides. When a high frequency (100 kHz-10MHz range) ultrasonic beam is directed at a gas-liquid interface, a geyser is formed and the height of the geyser is proportional to the acoustic intensity. Its formation is accompanied by the generation of a spray, resulting from the vibrations at the liquid surface and cavitation at the liquid-gas interface. The quantity of spray is a function of intensity. Ultrasonic atomization is accomplished by using an appropriate transducer made of PZT located at the bottom of the liquid container. A 500-1000 kHz transducer is generally adequate. The atomized spray which goes up in a column (fixed to the liquid container) is deposited on a suitable solid substrate and then heat-treated to obtain the film of the concerned material. The flow rate of the spray is controlled by the flow rate of air or any other gas. The liquid is heated to some extent, but its vaporization should be avoided. In Fig. 12.1 we show the apparatus employed in this method.

The source liquid would contain the relevant cations in the form of salts dissolved in an organic solvent. Organometallic compounds (e.g. acetates, alkoxides, β-diketonates etc.) are generally used for the purpose. Proper gas flow is crucial to obtain satisfactory conditions for obtaining a good spray. The pyrosol process is somewhere between CVD and spray pyrolysis, but the choice of source compounds for the pyrosol process is much larger than available for CVD. Furthermore, the use of a solvent minimizes or eliminates the difficulties faced in MOCVD. Films of a variety of materials have been obtained by the pyrosol method. The pyrosol method is truly inexpensive

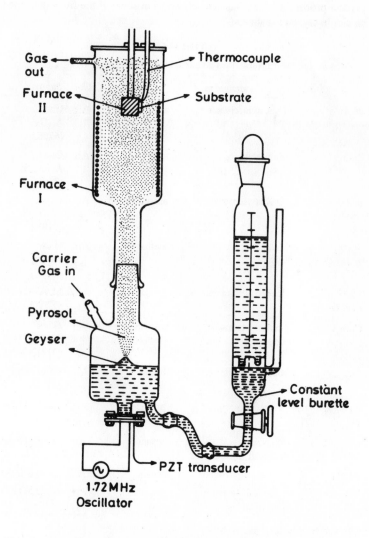

Fig. 12.1 Apparatus employed for preparing films by nebulized spray pyrolysis (From Raju and Rao).

compared to CVD/MOCVD. The thickness of pyrosol films can be anywhere between a few hundred angstroms to a few microns.

In Table 12.1 we list typical materials prepared by the method. Films of superconducting cuprates such $YBa_2Cu_3O_7$ have been prepared by the pyrosol process. Epitaxy has been observed in some of the films deposited on single-crystal substrates.

Table 12.1

Typical films prepared by pyrosol process

Material	Compound used	Solvent	Gas	Substrate (a)
Pt	Pt acetylacetonate	acetylacetone	Air	Glass, Al_2O_3, Si (670K)
ZnO	Zn acetate	methanol	Air	Glass, Al_2O_3, Si (770K)
In_2O_3	In acetylacetonate	acetylacetone	Air	Glass, Al_2O_3, Si (770K)
SnO_2	$SnCl_4$	methanol	Air, N_2	Glass, Al_2O_3, Si (670K)
$La_4Ni_3O_{10}$, $LaNiO_3$	La+Ni acetylacetonates	ethanol	Air/ O_2	Si, Al_2O_3 (770K)
$CdIn_2O_4$	In acetylacetonate, + Cd acetate	acetylacetone, methanol	Air	Glass, Al_2O_3, (710K)
TiO_2	butyl-orthotitanate	butanol, acetylacetone	Air, N_2	Glass, steel (670K)
γ-Fe_2O_3	Fe acetylacetonate	butanol	Air/ Argon	Glass, Al_2O_3 (760K)
(Ni, Zn) Fe_2O_4	Ni, Zn, Fe acetylacetonates	butanol	Air	Glass (770K)
Al_2O_3	Al isopropoxide	butanol	Air	Glass (920K)
$YBa_2Cu_3O_7$	dipivaloylmethane derivatives	butanol	Air/ O_2	MgO, $SrTiO_3$ (870K)

(a)   substrate temperature is shown.

References

1. M. Langlett and J.C. Joubert in *Chemistry of Advanced Materials* (C.N.R. Rao, ed), Blackwell, Oxford, 1992.

13

Arc and Skull Methods

The electric arc is conveniently used for the preparation of materials as well as for the growth of crystals of refractory solids [1, 2]. An arc for synthetic purposes is produced by passing a high current from a tungsten cathode to a crucible anode which acts as the container for the material to be synthesized (Fig. 13.1). The cathode tip is ground to a point in order to sustain a high current density. Typical operating conditions involve currents of the order of 70 amp at 15 volts. The arc is maintained in inert (He, Ar, N_2) or reducing (H_2) atmospheres. Even traces of oxygen attack the tungsten electrode and the gases are therefore freed from oxygen (by gettering with heated titanium sponge) before passing them into the arc chamber. The arc can be maintained in an oxygen atmosphere using graphite electrodes instead of tungsten. The crucible (anode) is made of a cylindrical copper block and is water-cooled during operation.

In order to synthesize materials, the starting materials are placed in the copper crucible. An arc is struck by allowing the cathode to touch the anode.

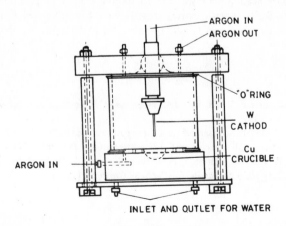

Fig. 13.1 DC Arc furnace

The current is slowly increased while the cathode is simultaneously withdrawn so as to maintain the arc. The arc is then positioned so that it bathes the sample in the crucible and the current increased until the reactants melt. When the arc is turned off, the product solidifies in the form of a button. Because of the enormous temperature gradient between the melt and the water-cooled crucible, a thin solid layer of the sample usually separates the melt from the copper hearth; in this sense, the sample forms its own crucible and hence contamination with copper does not take place. Contamination of the sample by tungsten vaporizing from the cathode can be avoided by using water-cooled cathodes. The arc method has been used for the synthesis of various oxides of Ti, V and Nb. Lower-valence rare earth oxides, $LnO_{1.5-x}$ have been prepared by arc fusion of Ln_2O_3 with Ln metal.

Skull melting is useful for preparing metal oxides as well as for growing crystals of these oxides [3]. The technique involves coupling of the material to a radio frequency electromagnetic field (200 kHz-4 MHz, 20-50 kW). The material is placed in a container consisting of a set of water-cooled cold fingers set in a water-cooled base (all made of copper), the space between the fingers being large enough to permit penetration of the electromagnetic field into the interior, but small enough to avoid leakage of the melt. The process is crucible-less and a thin solid skull separates the melt from the water-cooled container. Large single crystals of oxides can be grown by this method and the mass of the starting materials can be up to 1 kg. Temperatures up to 3600K are reached in this technique, permitting growth of crystals of materials like ThO_2 and stabilized ZrO_2. The stoichiometry of the oxide is readily controlled by the use of an appropriate ambient gas (CO/CO_2 mixtures, air or oxygen). Large crystals of CoO, MnO and Fe_3O_4 have been grown by the skull method. In CoO and MnO, trivalent metal ions were eliminated by heating the crystals in an appropriate CO/CO_2 mixture (see Section 2). Stoichiometric Fe_3O_4 crystals have been prepared similarly. Crystals of La_2NiO_4 and Nd_2NiO_4 have also been grown by the skull method. [4]

References

1. R.E. Loehman, C.N.R. Rao, J.M. Honig and C.E. Smith, *J. Sci. Indstr. Res.*, **28** (1969) 13.
2. H.K. Müller-Buschbaum, *Angew. Chem. Int. Ed.* (Engl) **20** (1981) 22.
3. H.R. Harrison, R. Aragon and C.J. Sandberg, *Mater. Res. Bull.* **15** (1980) 571.
4. C.N.R. Rao, D.J. Buttrey, N. Otsuka, P. Ganguly, H.R. Harrison, C.J. Sandberg and J.M. Honig, *J. Solid State Chem*, **51** (1984) 266.

14

Reactions at High Pressures

High pressure synthesis of solids has become common in recent years. Today, commercial equipment permitting simultaneous use of both high pressures and high temperatures is available. There are reviews of experimental aspects of high-pressure techniques in the literature [1-5]. For the 1-10 kbar pressure range, the hydrothermal method is often employed. Pressures in the range 10-150 kbar are used commonly for solid-state synthesis. The piston-cylinder apparatus (Fig. 14.1) employed for synthesis in this pressure range consists of a tungsten carbide chamber and a piston assembly. The sample is contained in a suitable metal capsule surrounded by a pressure-transducer (pyrophyllite). Pressure is generated by moving the piston through the blind hole in the cylinder. A microfurnace made of graphite or molybdenum is incorporated in the design. Pressures up to 50 kbar and temperatures up to 1800 K are readily reached in a volume of 0.1 cm^3 using this design. In the anvil apparatus (Fig. 14.2), the sample is subjected to pressure by simply squeezing it between two opposed anvils. Although pressures of ~200 kbar and temperatures up to 1300 K are reached

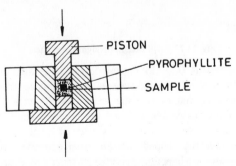

Fig. 14.1   Piston-cylinder apparatus

in this technique, it is not popular for solid-state synthesis since only milligram quantities can be handled. An extension of the opposed anvil principle (Fig. 14.2a) is the tetrahedral anvil design (Fig. 14.2b), where four massively supported anvils disposed tetrahedrally ram towards the centre where the sample is located in a pyrophyllite medium together with a heating arrangement. The multi-anvil design has been extended to cubic geometry, where six anvils act on the faces of a pyrophyllite cube located at the centre (Fig. 14.2c).

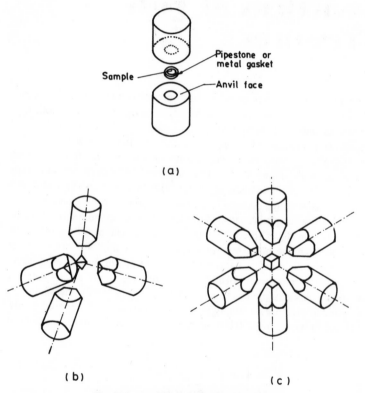

(a)

(b)   (c)

Fig. 14.2   Anvil designs (Bridgman and after)

The belt apparatus (Fig. 14.3) provides an ideal high pressure-high temperature combination for solid-state synthesis. This apparatus was used for the synthesis of diamonds some years ago. It actually involves a combination of the piston-cylinder and the opposed anvil designs. The apparatus consists of two conical pistons made of tungsten carbide which ram through a specially shaped chamber from opposite directions. The chamber and pistons are laterally supported by several steel rings making it possible routinely to reach fairly high pressures (~150 kbar) and high temperatures (~2300 K). In the belt apparatus, the sample is contained in a noble metal capsule (a BN or MgO container is used for chalcogenides) and surrounded by pyrophyllite and a graphite sleeve, the latter serving as an internal heater. In a typical high-pressure run, the sample is loaded, the pressure raised to the desired value and then the temperature increased. After holding the pressure for about 30 minutes, the sample is quenched (400 Ks^{-1}) while still under pressure. The pressure is released after the sample has cooled to room temperature.

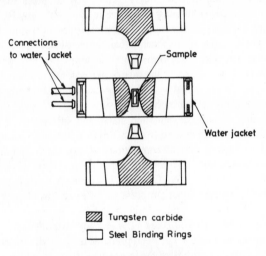

Fig. 14.3   Belt apparatus

High-pressure methods are generally used for the synthesis of materials that cannot otherwise be made. The formation of a new compound from its components requires that the new composition have a lower free energy that the sum of the free energies of the components. Pressure aids in the lowering of free energy in different ways [3]: (a). Pressure delocalizes outer d electrons in transition-metal compounds by increasing the magnitude of coupling between the d electrons on neighbouring cations, thereby lowering the free energy. A typical example is the synthesis of $ACrO_3$ (A = Ca, Sr, Pb) pervoskites and CrO_2. (b) Pressure stabilizes higher oxidation states of transition metals, thus promoting the formation of a new phase. For example, in the Ca-Fe-O system, only $CaFeO_{2.5}$ (brownmillerite) is stable under ambient pressures. Under high oxygen pressures, iron is oxidized to the 4^+ state and hence $CaFeO_3$ with the pervoskite structure is formed. (c) Pressure suppresses the ferroelectric displacement of cations thereby permitting the synthesis of new phases. The synthesis of A_xMoO_3 bronzes, for example, requires populating the empty d orbitals centred on molybdenum; at ambient pressures, MoO_3 is stabilized by a ferroelectric distortion of MoO_6 octahedra up to the melting point. (d) Pressure alters site-preference energies of cations, and facilitates the formation of a new phase. For example, it is not possible to synthesize $A^{2+}Mn^{4+}O_3$ (A = Mg, Co, Zn) ilmenites because of the strong tetrahedral site preference of the divalent cations. One therefore obtains a mixture of $A[AMn]O_4$ (spinel) + MnO_2 (rutile) under atmospheric pressure, instead of monophasic $AMnO_3$. However, the latter is formed at high pressures with a corundum-type structure in which both the A and Mn ions are in octahedral coordination. (e) Pressure can suppress the $6s^2$ core polarization in oxides containing isoelectronic Tl^{1+}, Pb^{2+} and Bi^{3+} cations. For example, $PbSnO_3$ cannot be made at atmospheric pressure because a mixture of $PbO + SnO_2$ is more stable than the perovskite. (f) Pressure can induce crystal structure transformations, the high-pressure phase being generally more close-packed. In the case of perovskites, pressure lowers the tolerance factor. In many instances, high-pressure phases can be quenched to retain the structures at atmospheric pressure.

Stabilization of unusual oxidation states and spin states of transition metals is of considerable interest (e.g. $La_2Pd_2O_7$). Such stabilization can be rationalized by making use of correlations of structural factors with the electronic configuration. Six-coordinated high-spin iron (IV) has been stabilized in $La_{1.5}Sr_{0.5}Li_{0.5}Fe_{0.5}O_4$ which has the K_2NiF_4 structure [6]. The elongated FeO_6 octahedra and the presence of ionic Li-O bonds resulting from the K_2NiF_4 structure favour the high-spin Fe (IV) state. The Li and Fe ions in this oxide are ordered in the ab-plane. Such an oxide can be prepared under oxidizing conditions. La_2LiFeO_6 prepared under high oxygen pressure has the perovskite structure with the iron in the pentavalent state. $CaFeO_3$ and $SrFeO_3$ prepared under oxygen pressure also contain octahedral Fe (IV); while Fe (IV) in $SrFeO_3$ is in the high-spin state with the e_g electron in the narrow σ^* band down to 4K, Fe (IV) in $CaFeO_3$ disproportionates to Fe (III) and Fe (V) below 290 K [7].

$LaNiO_3$ with Ni in the 3+ state can be prepared at atmospheric pressure; other rare earth nickelates have been prepared at high oxygen pressures. Recently $NdNiO_3$ has been prepared by sol-gel and other chemical routes [8, 9]. $MNiO_{3-x}$ (M = Ba or Sr) prepared under high pressure contain Ni (IV) [10]. In $La_2Li_{0.5}Co_{0.5}O_4$, the low-spin Co (III) ions transform to the intermediate- as well as high-spin states. The Li and Co ions are ordered in the ab-plane of this oxide of K_2NiF_4 structure; the highly elongated CoO_6 octahedra seem to stabilize the intermediate-spin state. Oxides in perovskite and K_2NiF_4 structures with trivalent Cu have been prepared under high oxygen pressure [6]. High F_2 pressure has been employed to prepare Cs_2NiF_6 and other fluorides [11]. Monel autoclaves are used in such reactions of F_2.

Solid state reactions can be quite slow under ordinary pressures even though the product is thermodynamically stable. Pressure has a marked effect on the kinetics of reactions, reducing reaction times considerably, and at the same time giving more homogeneous and crystalline products. For instance, $LnFeO_3$, $LnRhO_3$ and $LnNiO_3$ (Ln = rare earth) are prepared in a matter of hours under high pressure - high temperature conditions, whereas at ambient pressure, the reactions require several days ($LnFeO_3$ and $LnRhO_3$) or they do not occur at all. In several (AX) $(ABX_3)_n$ series of compounds, the end members ABX_3 and A_2BX_4, having the perovskite and K_2NiF_4 structures respectively, are formed at atmospheric pressure but not the intermediate phases such as $A_3B_2X_7$ and $A_4B_3X_{10}$. Pressure facilitates the synthesis of such solids; $Sr_3Ru_2O_7$ and $Sr_4Ru_3O_{10}$ are formed in 15 min at 20 kbar and 1300K. TaS_3, $NbSe_3$ and such solids can be prepared in 30 mins at 2 GPa and 970K.

High pressures have been employed for the synthesis of certain superconducting cuprates [13]. A simple example is the preparation of oxygen-excess superconducting La_2CuO_4 under high oxygen pressure. A more interesting example is the synthesis of the next homologue with two CuO_2 layers. $La_2Ca_{1-x}Sr_xCu_2O_6$ which had earlier been found to be an insulator was rendered superconducting by heating it under oxygen pressure [14]. $YBa_2Cu_4O_8$ was first prepared under high oxygen pressure. but this was soon found unnecessary [15]. However, superconducting cuprates with infinite CuO_2 layers of the type $Ca_{1-x}Sr_xCuO_2$ or $Sr_{1-x}Nd_xCuO_2$ can only be prepared under high hydrostatic pressure which help to give materials with shorter Cu-O bonds [16, 17]. It should be noted that $Ca_{1-x}Sr_xCuO_2$ prepared at ambient pressure is an insulator.

Hydrothermal Synthesis: In the hydrothermal method, the reaction is carried out either in an open or a closed system. In the open system, the solid is in direct contact with the reacting gas (F_2, O_2 or N_2) which also serves as a pressure-intensifier. A gold container is generally used in this type of synthesis. This method has been used for the synthesis of transition metal compounds (e.g. RhO_2, PtO_2 and Na_2NiF_6) with the transition metal in a high oxidation state. In Fig. 14.4 we show typical hydrothermal reactors. Hydrothermal high-pressure synthesis under closed system conditions has been employed for the preparation of higher-valence metal oxides. An internal oxidant such as $KClO_3$ is added to the reactants, which on decomposition under reaction conditions provides the necessary oxygen pressure. Pyrochlores of palladium (IV) and platinum (IV) of the type $Ln_2M_2O_7$ (Ln = rare earth) are typical of the solids prepared by this method (970K, 3 kbar). $(H_3O) Zr_2 (PO_4)_3$ and a family of zero thermal expansion ceramics (e.g. $Ca_{0.5}Ti_2P_3O_{12}$) have been prepared hydrothermally [18, 19]. Another good example is the synthesis of borates of Al, Y and such metals wherein the sesquioxides are reacted with boric acid [5]. Oxyfluorides have been prepared in HF medium [20].

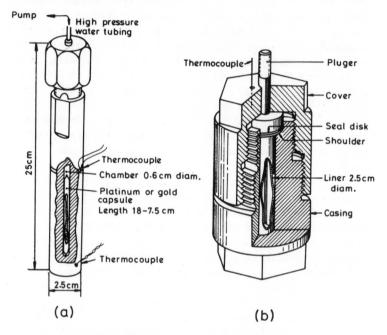

Fig. 14.4   Hydrothermal reactors: (a) typical reactor ; (b) Morey type reactor

Zeolites are generally prepared under hydrothermal conditions in the presence of alkali [21-23]. The alkali, the silica component and the source of aluminium are mixed in appropriate proportions and heated. The reactant mixture forms a hydrous gel which is then allowed to crystallize under pressure for several hours to several weeks between 330 and 470K. In a typical synthesis, $Al_2O_3.3H_2O$ dissolved in concentrated NaOH solution (20N) is mixed with a 1N solution of $Na_2SiO_3.9H_2O$ to obtain a gel (of composition 2.1 $Na_2O.Al_2O_3$. 2.l$SiO_2.6OH_2O$) which is then crystallized to give zeolite A. The $Na_2O-SiO_2-Al_2O_3-H_2O$ system yields a large number of materials with the zeolitic framework. Under alkaline conditions, Al is present as $Al(OH)_4$ anions. The OH$^-$ ion acts as a mineralizing catalyst while the cations present in the reactant mixture determine the kinds of zeolite formed. Besides water, some inorganic salts are also encapsulated in some zeolites. Several zeolite structures are found in the $K_2O-SiO_2-Al_2O_3-H_2O$ system as well. Li_2O however does not give rise to many microporous materials. Group II A cations yield several zeolitic products.

Zeolitization in the presence of organic bases is useful for synthesizing silica-rich zeolites. Silicalite with a tetrahedral framework enclosing a three-dimensional system of channels (defined by 10 rings wide enough to absorb molecules up to 0.6 nm in diameter) has been synthesized by the reaction of tetrapropylammonium (TPA) hydroxide and a reactive form of silica between 370 and 470K. The precursor crystals have the composition $(TPA)_2$ $0.48SiO_2.H_2O$ and the organic cation is removed by chemical reaction or thermal decomposition to yield microporous silicalite which may be considered to be a new polymorph of SiO_2 [23]. The clathrasil (silica analogue of a gas hydrate), dodecasil-1H, is prepared from an aqueous solution of tetramethoxy- silane and $N(CH_3)_4OH$; after the addition of aminoadamantane, the solution is treated hydrothermally under nitrogen for four days at 470K [24]. Fluoride ions act as good mineralizers in the synthesis of porous alumninosilicates and get trapped in the smallest cages [24a].

The use of template cations has enabled the synthesis of a variety of zeolite materials. Cations such as $(NMe_4)^+$ fit snugly into the cages (e.g. sodalite cages of sodalite and SAPO or gmelinite cages of zeolites omega). Neutral organic amines have also been used (e.g. in the synthesis of ZSM–5). Many new microporous materials including those based on $AlPO_4$ (analogue of SiO_2), gallosilicates and aluminogermanates (analogues of aluminosilicates), have been prepared. $AlPO_4$–based materials are prepared by the crystallization of gels formed by adding an organic template to a mixture of active alumina, H_3PO_4 and water at a pH of 5-8 around 470K. Metal aluminium phosphates (MAPO) with Mg, Ni, Zn and such divalent cations have been prepared similarly; the Mg derivative is a highly acidic catalyst. Recently, a novel large-pore microporous Mg-conaining aluminium phosphate (DAF–1) with a tetrahedral coordinated framework possessing two parallel channel systems with circular aperture openings of 0.61 and 0.75 nm (the latter containing supercages 0.163 nm in dia) has been synthesized [25]. This is prepared starting with MgO, Al_2O_3, P_2O_5, H_2O, acetic acid and decamethonium hydroxide in proper proportions.

One of the important recent developments in the synthesis of porous materials is that the pore sizes can be varied continuously between 20 Å to greater than 100 Å by using lyotropic molecules as templates for inorganic condensation [26, 27]. The pore wall structure, the mechanism of synthesis and the chemistry of these materials show how cooperative templating involves dynamic self-organization of inorganic and micellar array phases. Synthetic parameters for aluminosilicate and other framework compositions which determine the formation of isotropic, lamellar, hexagonal and cubic mesopore phase have been investigated in some detail recently. The work of Stucky in this area is noteworthy.

The closed shell nature of aluminosilicates renders them ineffective for certain reactions favoured by transition (*d*-block) elements. Haushalter has made efforts to prepare stable shape-selective micro- porous solids involving molybdenum phosphates [28]. These solids are prepared hydrothermally in aqueous H_3PO_4 in the presence of cationic templates along with anionic octahedral-tetrahedral frameworks containing Mo in oxidation state less than 5+ and possessing Mo-Mo bonds. Some of these contain around 40 vol% accessible internal void space. There is rich chemistry in these systems and there is considerable potential for applications. Based on this approach one may indeed discover novel microporous and catalytic oxide systems.

Some of the materials investigated by Haushalter are:

$(Et_4N)_6 Na [Na_{12} (H_3PO_4) \{Mo_6O_{15} (HPO_4)_3\}]. xH_2O,$

$(Et_4N)_2 [Mo_4O_8(PO_4)_{2/2} (H_{1.5}PO_4)_2]. 2H_2O,$

$Na_3 [Mo_2O_4 (HPO_4) (PO_4)]c. 2H_2O,$

$Pr_4N(NH_4) [Mo_4O_8 (PO_4)_2],$

$CH_3NH_3 [Mo_2O_2 (PO_4)_2 (H_2PO_4)]$

and $Cs(H_3O) [Mo_2O_2 (PO_4)_2 (HPO_4)]$

References

1. J.D. Corbett in *Solid State Chemistry - Techniques* (A.K. Cheetham and P. Day, eds), Clarendon Press, Oxford, 1987.
2. C.N.R. Rao and J. Gopalakrishnan, *New Directions in Solid State Chemistry*, Cambridge University Press, 1989.
3. J.B. Goodenough, J.A. Kafalas and J.M. Longo in *Preparative Methods in Solid State Chemistry* (P. Hagenmuller, ed), Academic Press, New York, 1972.
4. C.W.F.T. Pistorious, *Progr. Solid State Chem.* **11** (1976) 1.
5. J.C. Joubert and J. Chenavas in *Treatise in Solid State Chemistry* (N.B. Hannay, ed.), Vol. 5, Plenum Press, New York, 1975.
6. G. Demazeau, B. Buffat, M. Pouchard and P. Hagenmuller, *J. Solid State Chem.* **45** (1982) 881; also *Z. Anorg. Allgem. Chem.* **491** (1982) 60.
7. M. Takano and Y. Takeda, *Bull. Inst. Chem. Res.* (Kyoto University, Japan) **61** (1983) 406.
8. C.N.R. Rao and Gopalakrishnan, *Acc. Chem. Res.*, **20** (1987) 20.
9. J.K. Vassilou, M. Hornbostel, R. Ziebarth and F.J. Disalvo, *J. Solid State Chem.* **81** (1989) 208.
10. Y. Takeda, F. Kanamaru, M. Shimada and M. Koizumi, *Acta Cryst.* **B32** (1976) 2464.
11. P. Hagenmuller (ed), *Solid Inorganic Fluorides,* Academic Press, New York, 1985.
12. F. Sugawara, Y. Syono and S. Akimoto, *Mater. Res. Bull.* **3** (1968) 529.
13. C.N.R. Rao, R. Nagarajan and R. Vijayaraghavan *Supercond. Sci. Tech.* **6** (1993) 1.

14. R.J. Cava, R. Batlogg, L.F. Schneemeyer and others, *Nature*, **345** (1990) 602.
15. C.N.R. Rao, G.N. Subbanna, R. Nagarajan and others, *J. Solid State Chem.*, **88** (1990) 163.
16. M. Azuma, M. Takano and others, *Nature*, 356 (1992) 775.
17. M. Takano, Z. Hiroi, M. Azuma and Y. Takeda in *Chemistry of High-Temperature Superconductors* (C.N.R. Rao, ed), World Scientific, 1992.
18. R. Roy, D.K. Agarwal, J. Alamo and R.A. Roy, *Mater. Res. Bull.* **19** (1984) 471.
19. M.A. Subramanian, B.D. Roberts and A. Clearfield, *Mater. Res. Bull.* **19** (1984) 1471.
20. A.W. Sleight, *Inorg. Chem.* **8** (1969) 1764.
21. J.M. Newsam in *Solid State Chemistry - Compounds* (A.K. Cheetham and P. Day, eds), Clarendon Press, Oxford, 1992.
22. R.M. Barrer, *Zeolites*, **1** (1981) 130.
23. E.M. Flanigen, J.M. Bennett, R.W. Grose, J.P. Cohen, R.L. Patton, R.M. Kirchner and J.V. Smith, *Nature*, **271** (1978) 512.
24. E.J.J. Groenen, N.C.M. Alma, A.G.T.M. Bastein, G.R. Hays, R. Huis and A.G.T.G. Kortbeck, *J. Chem. Soc. Chem. Commun.* (1983), 1360.
24a M. Estermann, L.B. McCusker, C. Baerlocher, A. Merrouche and H. Kessler, *Nature*, 352 (1991) 320.
25. P.A. Wright, R.H. Jones, S. Natarajan, R.G. Bell, J. Chen, M.B. Hursthouse and J.M. Thomas, *J. Chem. Soc. Chem. Commun.* 633 (1993).
26. J. Beck, C.T. Kresge, M.E. Leonowicz, W.J. Roth, J.C. Vartuli, C.T.W. Chu and I.D. Johnson, US Patent 5, 057, 296 (Mobil R&D).
27. J. Beck et. al *J. Am. Chem. Soc.*, **114** (1992) 10834.
28. R.C. Haushalter and L.A. Mundi, *Chem. Mater.* **4** (1992) 31.

15

Intergrowth Structures

There are several metal oxides exhibiting well-defined recurrent intergrowth structures with large periodicities, rather than forming random solid solutions with variable composition. Such ordered intergrowth structures themselves however frequently show the presence of wrong sequences. The presence of wrong sequences or lamellae is best revealed by a technique that is more suited to the study of local structure. High resolution electron microscopy (HREM) enables a direct examination of the extent to which a particular ordered arrangement repeats itself and the presence of different sequences of intergrowths, often of unit cell dimensions. Selected area electron diffraction which forms an essential part of HREM provides useful information (not generally provided by x-ray diffraction) regarding the presence of supercells due to the formation of intergrowth structures. Many systems forming ordered intergrowth structures have come to be known in recent years [1, 2]. These systems generally exhibit homology. In Table 15.1 various known intergrowth structures are listed. What is amazing is that such periodicity occurs in three-dimensional solids routinely prepared by ceramic procedures. The factors responsible for such order are not fully clear.

If the ABO_3 perovskite structure is cut parallel to the (110) plane, slabs of the compositions $A_{n-1}B_nO_{3n+2}$ are obtained; if these slabs are stacked, an extra sheet of A gets introduced giving rise to the family of oxides of the general formula $A_nB_nO_{3n+2}$. Typical members of this family are $Ca_2Nb_2O_7$ ($n=4$), $NaCa_4Nb_5O_{17}$ ($n=5$) and $Na_2Ca_4Nb_6O_{20}$ ($n=6$). HREM and x-ray diffraction show that an ordered intergrowth structure with $n=4.5$, with the composition $NaCa_8Nb_9O_{31}$, corresponds to alternate stacking of $n=4$ and $n=5$ lamellae. What is curious is that $NaCa_8Nb_9O_{31}$ is prepared by the standard procedure of heating the mixture of component oxides and yet shows such extraordinary periodicity. Between $n=4$ and 4.5, a large number of ordered solids are found with the b parameter of the unit cell ranging anywhere from 58.6Å in the $n=4.5$ compound to a few thousand angstroms in longer period structures. These solids seem to belong to a class of infinitely adaptive structures, fist envisaged by Anderson [3].

Table 15.1

Ordered intergrowth structures forming homologus series

1. Barium ferrites

    (i) MpYq where M = $BaFe_{12}O_{19}$ and Y = $BaMe_2Fe_{12}O_{22}$ (Me = Zn, Ni etc.); $Ba_{2n+p}Me_{2n}Fe_{12\,(n+p)}O_{22n+19p}$ with n = 1 to 47 and p = 2 to 12.

    (ii) MS_n Where S = $Me_2Fe_4O_8$ with n = 1, 2, 3, 4.

2. Perovskites

    (i) $Bi_4A_{m+n-2}B_{m+n}O_{3\,(m+n)+6}$ formed by Aurivilluius oxides of the type $Bi_2A_{n-1}B_nO_{3n+3}$.

    (ii) $A_nB_nO_{3n+1}$ such as $(Na, Ca)_nNb_nO_{3n+2}$ with n = 4 - 4. 5.

    (iii) $A_{n+1}B_nO_{3n+1}$ as exemplified by Sr-Ti-O and La-Ni-O systems.

3. Tungsten bronzes

    (i) A_xWO_3 (ITB) with A = Alkali metal, Bi etc.

    (ii) $A_xM_xW_{1-x}O_3$ (bronzoids) with M = V, Nb etc.

    (ii) $A_xP_4O_8(WO_3)_{2m}$ with m = 4 - 10.

    (iv) $K_xP_2O_4(WO_3)_{2m}$ with m = 2.

    (v) $P_4O_8(WO_3)_{4m}$.

    (vi) $ACu_3M_7O_{21}$ with M = Ta, Nb.

4. Siliconiobates

    (i) $(A_3M_6Si_4O_{26})_n (A_3Nb_{8-x}M_xO_{21})$ with A = Ba, Sr; M = Ta, Nb.

    (ii) $(Ba_3Nb_6Si_4O_{26})_nA_3Nb_xM_xO_{21}$ with A = K or Ba; M = Ti, Ni, Zn, etc.

5. Others

    (i) $(ATi_6O_{13})_n (ATi_4O_9)_m$ with A = Na, A = Ba.

    (ii) La_2O_3 - ThO_2 system.

Aurivillius described the family of oxides of the general formula $Bi_2A_{n-1}B_nO_{3n+3}$ where the perovskite slabs, $(A_{n-1}B_nO_{3n+1})^{2-}$, n octahedra thick, are interleaved by $(Bi_2O_2)^{2+}$ layers. Typical members of this family are Bi_2WO_6 (n = 1), $Bi_3Ti_{1.5}W_{0.5}O_9$ (n = 2), $Bi_4Ti_3O_{12}$ (n = 3) and $Bi_5Ti_3CrO_{15}$ (n = 4). These oxides form intergrowth structure of the general formula $Bi_4A_{m+n-2}B_{m+n}O_{3(m+n)+6}$ involving alternate stacking of two Aurivillius oxides with different n values (Fig. 15.1). The method of preparation involves simply heating a mixture of the component metal oxides at ~1000 K. Ordered intergrowth structures with (m, n) values of (1, 2), (2, 3) and (3, 4) have been characterized [4] by HREM (see Fig. 15.2). It is intriguing that such intergrowth structures with long range order are indeed formed while either member (m and n) can exist as a stable entity. These materials seems to be truly representative or recurrent intergrowth. The periodicity found in recurrent intergrowth solids formed by the Aurivillius family of oxides is indeed remarkable.

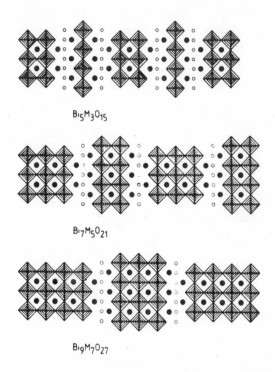

$Bi_5M_3O_{15}$

$Bi_7M_5O_{21}$

$Bi_9M_7O_{27}$

Fig. 15.1   Different types of intergrowth structures formed by the Aurivillius family of bismuth oxides. Notice the intergrowth of (1, 2), (2, 3) and (3, 4) layered units

WO_3 forms tetragonal, hexagonal, or perovskite-type bronzes of the general formula B_xWO_3 by the interaction with alkali and other metals. The family of intergrowth tungsten bronzes (ITB) involving the intergrowth of nWO_3 slabs and one to three strips of the hexagonal tungsten bronze (HTB) is of interest. In the intergrowth tungsten bronzes, x in M_xWO_3 is generally 0.1 or less. Depending on whether the HTB strip is one or two-tunnel wide, ITB's are classified as belonging to (0, n) and (1, n) series (Fig. 15.3).

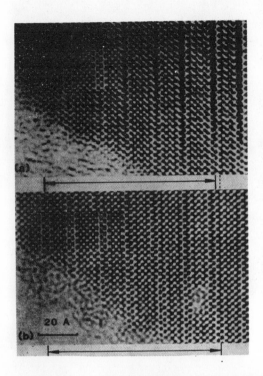

Fig. 15.2   High resolution electron micrograph of (3, 4) intergrowth structures:
(a) $Bi_9Ti_6CrO_{27}$ involving the Aruivillius phases $Bi_4Ti_3O_{12}$ (n = 3) $Bi_5Ti_3CrO_{15}$ (n = 4)
and (b) $BaBi_8Ti_7O_{27}$. Computer simulated images and unit cell lengths are shown (from
Jefferson et al *Mat. Res. Bull.* 1984)

Two-tunnel wide HTB strips seem to be most stable in ITB's and several ordered sequences of the (0, n) and the (1, n) series have been identified [5]. In the ITB phases of Bi, the HTB strips are always one-tunnel side (Fig. 15.4). Displacement of adjacent tunnel rows due to the tilting of WO_3 octahedra often results in the doubling of the long-period axis of the ITB. Evidence for the ordering of the intercalating Bi atoms in the tunnels has been found in terms of satellites around the superlattice spots in the electron diffraction patterns [6].

Among the other systems exhibiting ordered intergrowth, the family of hexagonal barium ferrites, M_pY_q (M=$BaFe_{12}O_{19}$ and Y = $Ba_2Me_{12}O_{22}$, where Me is Zn, Ni Mg. etc.) is noteworthy (Fig. 15.5). A number of intergrowth structures of this family have been identified [7] and they have all been prepared by ceramic procedures.

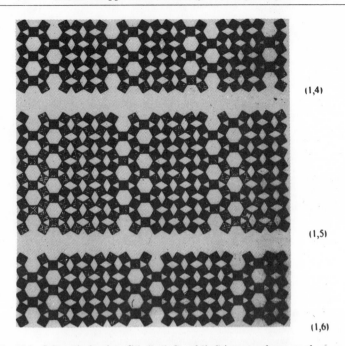

(1,4)

(1,5)

(1,6)

Fig. 15.3 Schematic drawing of (1, 4), (1, 5) and (1, 6) intergrowth tungsten bronzes. Hexagonal tunnels of HTB strips separate the WO$_3$ slabs shown in polyhedral from (after Kihlborg 1979)

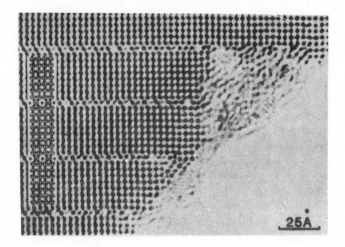

Fig. 15.4 HREM of Bi$_x$WO$_3$ intergrowth bronze. The dark circles between the WO$_3$ slabs represent Bi atoms (from Ramanan et al *Proc. Roy. Soc. Lond.* 1984)

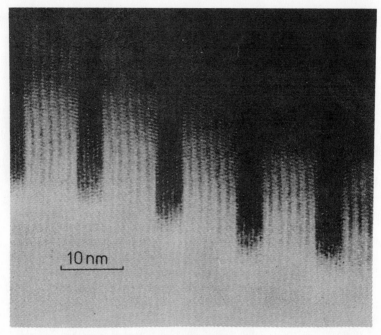

Fig. 15.5 HREM of $MYMY_6$ intergrowth in barium ferrite: $M = BaFe_{12}O_{19}$; $Y = Ba_2Me_2Fe_{12}O_{22}$ (Me = Zn, Ni or Mg) (From Anderson and Hutchison *Contemp. Phys.* 1975)

References

1. C.N.R. Rao, *Bull. Mater. Sci. 7* (1985) 155.
2. C.N.R. Rao and J.M. Thomas, *Acc. Chem. Res. 18* (1985) 113.
3. J.S. Anderson. *J. Chem. Soc. Dalton Trans.* (1993) 1107.
4. D.A. Jefferson, M.K. Uppal and C.N.R. Rao, *Mat. Res. Bull 19* (1984) 1403.
5. L. Kihlborg, Nobel Symposium 47, *Royal Swedish Acad. Sci.,* 1979.
6. A. Ramanan, J. Gopalakrishnan, M.K. Uppal, D.A. Jefferson and C.N.R. Rao, *Proc. Roy. Soc. London, A395* (1984) 127.
7. J.S. Anderson and J.L. Hutchison, *Contemp. Phys. 16* (1975) 443.

16

Superconducting Cuprates

Bednorz and Müller [1] discovered high T_c superconductivity (~30K) in $La_{2-x}Ba_xCuO_4$. The discovery of a superconducting cuprate with a T_c above 77K created sensation in early 1987. Wu et al [2] who announced this discovery first made measurements on a mixture of oxides containing Y, Ba and Cu obtained in their efforts to prepare the Y-analogue of $La_{2-x}Ba_xCuO_4$. Rao et al [3] independently worked on the Y–Ba–Cu–O system on the basis of an understanding of solid state chemistry. Rao et al knew that Y_2CuO_4 could not be made and that substituting Y by Ba in this cuprate was not the way to proceed. They therefore tried to make $Y_3Ba_3Cu_6O_{14}$ by analogy with $La_3Ba_3 Cu_6O_{14}$ described earlier by Raveau et al and varied the Y/Ba ratio as in $Y_{3-x}Ba_{3+x}Cu_6O_{14}$. By making $x = 1$, they obtained $YBa_2Cu_3O_7$ ($T_c \sim 90K$). They knew the structure had to be that of a defect perovskite from the beginning, because of the route adopted for the synthesis.

Synthesis of cuprate superconductors has been reviewed extensively by Rao et al [4]. We shall briefly examine some preparative aspects of these materials, although we have mentioned some of them under the different methods of synthesis discussed earlier. Cuprates are generally made by the traditional ceramic method which involves thoroughly mixing the various oxides, carbonates, oxalates of the component metals in the desired proportion and heating the mixture (in pellet form) at a high temperature. The mixture is ground again after some time and reheated until the desired product is formed as indicated by x-ray diffraction. This method does not always yield the product with the desired structure, purity or oxygen stoichiometry. Variants of this method have been employed. For example, decomposing a mixture of nitrates has been found to yield a better product in the case of the 123 compounds by some workers. Some workers have used BaO_2 in place of $BaCO_3$ for the synthesis.

One of the problems with the superconducting bismuth cuprates and thallium cuprates (Fig. 16.1) is the difficulty in obtaining phasic purity (minimizing the intergrowth of the different layered phases). The glass or the melt route has been employed to obtain better samples of bismuth cuprates. The method involves preparing a glass by quenching the melt; the glass is then crystalized by heating it above the crystallization temperature. Thallium cuprates are best prepared in sealed tubes (gold or silver). Heating Tl_2O_3 with a matrix of the other oxides (already heated to 1100–1200K) in a sealed tube is preferred by some workers. It is important that thallium cuprates are not prepared in open furnaces since Tl_2O_3 (which readily sublimes) is highly toxic. The same is true of mercury cuprates; sealed tube reaction is essential here since mercury can be formed by the decomposition of the oxide. In order to obtain superconducting compositions corresponding to a particular copper content (number of CuO_2 sheets) by the ceramic method, one often has to start with various arbitrary compositions especially in the case of the Tl cuprates. The real composition of a bismuth or a thallium cuprate superconductor may not be anywhere near the starting composition. The actual composition has to be determined by analytical election microscopy an other methods.

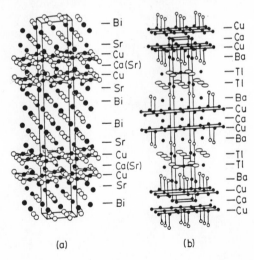

Fig. 16.1   Superconducting $Bi_2CaSr_2Cu_2O_8$ and $Tl_2 CaBa_2Cu_2O_8$.

Heating oxidic materials under high oxygen pressures or in flowing oxygen often becomes necessary to attain the desired oxygen stoichiometry. Thus, La_2CuO_4 and $La_2Ca_{1-x}Sr_xCu_2O_6$ heated under high oxygen pressures become superconducting with T_cs of 40 and 60K respectively. In the case of the 123 compounds, one of the problems is that they lose oxygen easily. Note that superconducting $LnBa_2Cu_3O_7$ (Ln = Y or rare earth) is orthorhombic while insulating $LnBa_2Cu_3O_6$ is tetragonal. It therefore becomes necessary to heat the material in an oxygen atmosphere below the orthorhombic-tetragonal transition temperature. Oxygen stoichiometry is not a serious problem with the bismuth cuprates. Many of the thallium cuprates (as prepared) tend to be oxygen-excess and show lower T_c values or do not exhibit super- conductivity. By annealing them in vacuum or in a hydrogen atmosphere, high T_cs are attained [4]. The real problem is to optimize the hole concentration by controlling oxygen stoichiometry.

The 124 superconductors (Fig.16.2) were first prepared under high oxygen pressures, but it was later found that heating the oxide or nitrate mixture in the presence of Na_2O_2 in flowing oxygen was sufficient to obtain 124 compounds. Analogues of $La_2Ca_{1-x}Sr_xCu_2O_6$ have been prepared by heating the mixture of oxides in the presence of $KClO_3$ [5]. Superconducting Pb cuprates, on the other hand, can only be prepared in the presence of very little oxygen (N_2 with a small percentage of O_2). The Cu^{1+} ions in these cuprates would otherwise get oxidized. In the case of the electron superconductor, $Nd_{2-x}Ce_xCuO_4$ (F16. 3), it is necessary to heat the material in an oxygen-deficient atmosphere; otherwise, the electron given by Ce will merely go into giving an oxygen-excess material. It may be best to prepare $Nd_{2-x}Ce_xCuO_4$ by a suitable method (say decomposi- tion of mixed oxalates or nitrates) and then reduce it with hydrogen.

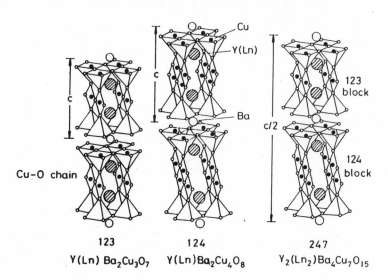

Fig. 16.2   Schematic structures of 123, 124 and 247 cuprates.

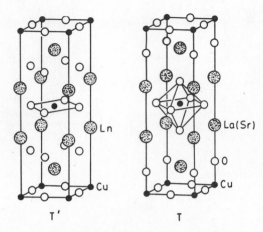

Fig. 16.3 $Nd_{2-x}Ce_xCuO_4(T')$ and $La_{2-x}Sr_xCuO_4$ (T)

The sol-gel technique has been effectively employed for the synthesis of 123 compounds such as $YBa_2Cu_3O_7$ and the bismuth cuprates. Materials prepared by such low-temperature methods have to be annealed or heated under suitable conditions to obtain the desired oxygen stoichiometry as well as the characteristic high T_c. 124 cuprates, lead cuprates and even thallium cuprates have been made by the sol-gel method; the first two are specially difficult to make by the ceramic method. Coprecipitation of all the cations in the form of a sparingly soluble salt such as carbonate or oxalate in a proper medium (e.g. using tetra-ethylammonium oxalate), followed by thermal decomposition of the dried precipitate has been employed by many workers to prepare cuprates.

Several other strategies have been employed for the synthesis of superconducting cuprates, as indicated in the earlier sections while discussing the different methods. Specially noteworthy are the use of the combustion method, the alkali-flux method and electrochemical oxidation for cuprate synthesis [4]. Superconducting infinite-layerd cuprates can be prepared only under high pressures because of bonding (structural) considerations [6]. Strategies where structural and bonding considerations are involved in the synthesis are generally more interesting. One such example is the synthesis of modulation-free superconducting bismuth cuprates [7]. Superconducting bismuth cuprates such as $Bi_2CaSr_2Cu_2O_8$ exhibit superlattice modulation. Since the modulation has something to do with the oxygen content in the Bi–O layers and lattice mismatch, one Bi^{3+} can be substituted by Pb^{2+} to eliminate the modulation, without losing superconductivity.

An important recent finding is that oxyanions such as carbonate and nitrate can replace copper in cuprate superconductors (Fig. 16.4). Generally CO_3^{2-} seems to partly replace square-planar CuO_4 units (e.g. Cu in the Cu–O

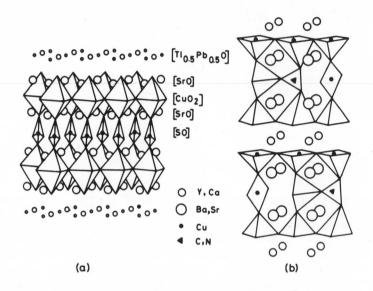

Fig. 16.4   Schematic structures of (a) $Tl_{0.5}Pb_{0.5}Sr_4Cu_2(SO_4)O_y$ and (b) $YCaBa_4(Ba_2Sr_2)Cu_5[CO_3]_{1-x}[NO_3]_xO_y$. SO_4 units are shown as tetrahedra while CO_3 and NO_3 units are shown by triangles.

chains of $YBa_2Cu_3O_7$). While carbonate destroys superconductivity in $YBa_2Cu_3O_7$, it has been possible to prepare a superconducting composition with the incorporation of NO_3^- ions along with Ca^{2+} (in Y^{3+} sites) in the 123 system. Part replacement of CuO_4 units by CO_3^{2-} converts the square-planar copper to a square-pyramidal ones [8]. Superconducting oxycarbonates of the bismuth and thallium cuprate families (e.g. $Bi_2Sr_4Cu_2CO_3O_8$ and $Tl_{0.5}Pb_{0.5}Sr_4Cu_2(CO_3)O_y$) as well as of the infinitely layered cuprate, $Ba_{2-x}Sr_xCuO_2(CO_3)$, prepared at 1270K under high pressure have been reported [9,10]. The possibility of having oxyanions as integral parts of the structure of oxides open up many possibilities. Phosphate and sulfate derivatives of cuprates have indeed been reported [11].

In Table 16.1 we list the various cuprate superconductors along with the T_c values and the preferred methods of synthesis.

Table 16.1

Synthesis of Superconducting Cuprates

Cuprate	*Approx T_c (K)*	*Methods of synthesis*[a]
$L_{2-x}Sr_x(Ba_x)CuO_4$	35	Ceramic[*]
$La_2Ca_{1-x}Sr_xCu_2O_6$	60	Ceramic (high O_2 Pressure)[*]
$La_2CuO_{2+\delta}$	35	Ceramic (high O_2 Pressure)[*], alkali flux
$YBa_2Cu_3O_7$ (b)	90	Ceramic (flowing O_2)[*], sol-gel[*], coprecipitation[*]
$YBa_2Cu_4O_8$	80	Ceramic (with Na_2O_2)[*]
$Bi_2CaSr_2Cu_2O_8$	90	Ceramic (air-quench)[*] sol-gel[*]
$Bi_2Ca_2Sr_2Cu_3O_{10}$	110	Ceramic[*], sol-gel, melt route coprecipitation
$TlCaBa_2Cu_2O_{6+\delta}$	90	Ceramic (sealed Ag/Au tube)[*]
$TlCa_2Ba_2Cu_3O_{8+\delta}$	115	Ceramic (sealed Ag/Au tube)[*]
$Tl_2CaBa_2Cu_3O_{8+\delta}$	110	Ceramic (sealed Ag/Au tube)[*]
$Tl_2Ca_2Ba_2Cu_3O_{10}$	125	Ceramic (sealed Ag/Au tube)[*]
$Tl_{0.5}Pb_{0.5}CaSr_2O_{6+\delta}$	110	Ceramic (sealed Ag/Au tube)[*]
$Hg_2Ba_2Ca_2Cu_3O_y$	133	Ceramic (sealed tube)[*]
$PbSr_2Ca_{1-x}Y_xCu_3O_8$	70	Ceramic (low O_2 partial pressure)[*] Sol-gel[*] (low O_2 partial pressure)
$Nd_{2-x}Ce_xCuO_4$	30	Ceramic (low O_2 partial pressure)[*] Coprecipitation (low O_2 partial pressure)[*]
$Ca_{1-x}Sr_xCuO_2$	40–110	Ceramic (high pressures)[*]
$Sr_{1-x}Nd_xCuO_2$	40–110	Ceramic (high pressures)[*]

(a) recommended method is indicated by an asterisk.

(b) other rare earth cuprates of the 123 type are prepared by similar methods.

References

1. J.G. Bednorz and K.A. Müller, *Z Physik* **B64** (1986) 189.
2. M.K. Wu et al, *Phys. Rev. lett.* **58** (1987) 908.
3. C.N.R. Rao, P. Ganguly, R.A. Mohan Ram, A.K. Raychaudhuri and K. Sreedhar, *Nature*, **326** (1987) 856.

4. C.N.R. Rao, R. Nagarajan and R. Vijayaraghavan, *Supercond. Sci. Tech.* **6** (1993) 1.

5. Unpublished results from the author's laboratory.

6. M. Azume, M. Takano et al, *Nature*, **356** (1992) 775. M. Takano, Z. Hiroi, M. Azume and Y. Takeda, in *Chemistry of High-Temperature Superconductors* (C.N.R. Rao, ed.), World Scientific, Singapore, 1992.

7. V. Manivannan, J. Gopalakrishnan and C.N.R. Rao, *Phys. Rev.* **B43** (1991) 8686.

8. A. Maignan, M. Hervieu C. Michel and B. Raveau, *Physica C,* **208** (1993) 116

9. M. Huve, C. Michel, A. Maignan, M. Hervieu, C. Martin and B. Raveau, *Physica C,* **205** (1993) 219.

10. K. Kinoshita and T. Yamoda, *Nature,* **357** (1992) 313.

11. S. Ayyappan, V. Manivannan, G.N. Subbanna and C.N.R. Rao, *Solid State Commun.*, **87** (1993) 551; R. Nagarajan, S. Ayyappan and C.N.R. Rao, *Physica C* **220** (1994) 373; C.N.R. Rao et al *Solid State Commun.*, **88**, 757 (1993).

17

Metal Borides, Carbides And Nitrides

Metal borides are generally prepared by the direct reaction of the elements at high temperatures or by the reduction of metal oxides or halides. Thus, reduction of mixtures of B_2O_3 and metal oxides by carbothermic reaction yield metal borides. Reaction of metal oxides with boron or with a mixture of carbon and boron carbide is another route. Some metal borides are prepared by fused salt electrolysis (e.g. TaB_2). Borides of IV A - VII A elements as well as ternary borides have been reviewed by Nowotny [1]. The method employed to prepare TiB_2 starting with $TiCl_4$ is interesting [2]. $TiCl_4$ and BCl_3 react with sodium in a nonpolar solvent (e.g. heptane) to produce an amorphous precursor powder along with NaCl. NaCl is distilled off and the precursor crystallized at relatively low temperatures ($\sim$970 K).

$$TiCl_4 + 2\,BCl_3 + 10\,Na \longrightarrow TiB_2\,(s) + 10\,NaCl\,(s)$$

$$TiB_2\,(s) + 10\,NaCl\,(g) \xrightarrow[970\,K]{vaccum} TiB_2\,(s) + 10\,NaCl\,(g)$$

The reaction probably proceeds through the formation of the $Cl_2B–TiCl_3$ intermediate.

Metal carbides are generally prepared by the direct reaction of the elements at high temperatures ($\sim$ 2470 K). Reaction of metal oxides with carbon is another important route. Reaction of metal vapour with hydrocarbons also yields metal carbides. Phase relations in carbides of IV A, V A and VI A group elements as well as actinides have been reviewed by Storms [3]. SiC has been prepared by the reaction of $SiCl_4$ and CCl_4 with Na; a similar reaction of CCl_4 and BCl_3 with Na gives B_4C [2]. SiC is formed by the decomposition of CH_3SiH_3 or $(CH_3)_2SiCl_2$. Pyrolysis of organosilicon polymer precursors has been employed to prepare SiC [4].

Some of the precursor reactions were discussed earlier in Section 4 of this monograph.

Metal nitrides are generally prepared by the direct reaction of the elements. Ionic nitrides are also prepared by the decomposition of metal amides as illustrated by the reaction,

$$3 \, Ba \, (NH_2)_2 \longrightarrow Ba_3N_2 + 4 \, NH_3$$

Transition metal nitrides where nitrogen is present as an interstitial are prepared by the reaction of the metals with NH_3 around 1470K. BN is obtained by heating boron with NH_3 at white heat. Metal nitrides are also prepared by the reaction of metal chlorides with NH_3. For example, AlN and WN films are prepared by the reaction of NH_3 with $AlCl_3$ and WCl_6 respectively. Recently, nitrides of Ti, Zr, Hf and lanthanides have been prepared by the reaction of lithium nitride with the anhydrous metal chlorides [5]:

$$4 \, Li_3N + 3 \, MCl_4 \longrightarrow 12 \, LiCl + 3 \, MN + \tfrac{1}{2} \, N_2$$

$$Li_3N + LnCl_3 \longrightarrow 3 \, LiCl + LnN$$

Phase relations in nitrides of IVA and VA elements and actinides have been reviewed [3]. GaN films have been prepared by the low-pressure CVD of HN_3 and $Ga(CH)_3$ by Flowers et al [6].

In antiperovskite nitrides of the type Mn_4N, Fe_4N and Fe_3PtN, nitrogen is present in the octahedral holes of the metal framework. Chern et al [7] have prepared a new antiperovskite nitride of the formula Ca_3MN where M is a group IV or V element. Here the A and B sites are occupied by M^{3-} and N^{3-} respectively and the anion sites by Ca^{2+}. These air-sensitive nitrides are prepared by grinding Ca_3N_2 (obtained by the reaction of Ca and N_2 at 1170K) with the third element and heating the pellet of the mixture at 1270K. Nitrides with M = P, As, Sb and Bi require lower temperatures than those with M = Ge, Sn and Pb. Ca_3CrN_3 has been prepared by the reaction of Cr_2N/CrN with Ca_3N_2 at 1620K for 4 days in a steel tube [8]. Starting from Sr_2N and Ba_3N_2, one obtains Sr_3CrN_3 and Ba_3CrN_3 respectively. Edwards and coworkers [9] have prepared $SrFeN_3$ by the ignition of strontium nitride in a sealed stainless steel capsule at 1370K. These workers have also prepared ternary lithium nitrides such as Li_5TiN_3 and Li_7VN_4 by the reaction of Li_3N with TiN and VN respectively at 1070K in a N_2 atmosphere. Li_3FeN_2 is obtained by the reaction of Li_3N with Fe. Bem and Zur Loye [10] have synthesized a new ternary nitride $FeWN_2$ by the ammonialysis of $FeWO_4$ at 1070 K. One can prepare ternary nitrides of the type M_3W_3N and MWN_2 (M = Co, Ni etc) by such a reaction. There is great scope for synthesising other ternary nitrides, specially those containing a transition metal (e.g. Co, Ni) in mixed valent state.

Silicon nitride ceramics have emerged to become important materials with many potential applications [11]. Reactive decomposition of $SiCl_4$ or

SiH_4 with NH_3 yields Si_3N_4. The use of precursors to prepare Si_3N_4 and other nitrides was discussed in Section 4 of the monogrph.

Oxynitrides of several metals have been prepared. Thus, oxynitrides of Zr in the $Zr_3N_4-ZrO_2$ system were prepared by the reaction of ZrN and ZrO_2 in NH_3 around 1370K [12]. A new cubic oxynitride of Zr is reported to be formed by the reaction of ZrO_2 and ZrN around 1870 in a $N_2 + H_2$ mixture [13]. Recently, Zr oxynitrides have been prepared by the reaction of ZrNCl and ZrO_2 at 1220K [14]. Silicon oxynitride ceramics, SIALONS, derived by the substitution of nitrogen in Si_3N_4 partly by oxygen and of Si by Al have been of considerable interest [11, 15]. Sialon powders have been prepared by the reaction of metakaolin with NH_3 vapour. Grins and cowokers have prepared baddeleyite type $Ta_{1-x}Zr_xO_{1+x}N_{1-x}$ ($0 < x < 1$) by the reaction of dried $TaCl_5-Zr$ propoxide gels with NH_3 vapour.

We give below the reactions involved in the conventional synthesis of Si_3N_4, SIALON, silicon oxynitride, AlN and BN:

(a) $3\,Si + 2\,N_2 \longrightarrow Si_3N_4$ (around 1220 K)

This reaction can be accomplished by simple heating, heating in a plasma or by the combustion method.

(b) $3\,SiCl_4 + 4\,NH_3 \longrightarrow Si_3N_4 + 12\,HCl$

$3\,SiH_4 + 4\,NH_3 \longrightarrow Si_3N_4 + 12\,H_2$

The above two reactions occur at high temperatures compared to the procedure involving the decomposition of the polymeric silicon diimide (obtained by the reaction of $SiCl_4$ with excess NH_3) or by the decomposition of the precursors discussed earlier in Section 4.

(c) $3\,SiO_2 + 6\,C + 2\,N_2 \longrightarrow Si_3N_4 + 6\,CO$ (~1720 K)

$3\,(Al_2O_3 . 2\,SiO_2) + 15\,C + 5\,N_2 \longrightarrow 2\,Si_3\,Al_3\,O_3\,N_5$
$\qquad\qquad\qquad\qquad\qquad\qquad\qquad\qquad + 15\,CO$

The above two reactions involve carbothermal reduction and nitridation. Ramesh and Rao [16] find that the carbothermic reaction of SiO_2 occurs readily if amorphous SiO_2 prepared by the oxidation of the commercial $SiO_{1.7}$ is used as the starting material.

(d) $Si_3N_4 + SiO_2 + 2AlN \longrightarrow Si_4Al_2O_2N_6$

This reaction corresponds to the process involved in reaction sintering.

(e) $3\,Si + SiO_2 + 2\,N_2 \longrightarrow 2\,Si_2\,N_2O$

$Si_3N_4 + SiO_2 \longrightarrow 2\,Si_2N_2O$

Silicon oxynitride, Si_2N_2O, has not been adequately investigated in terms of applications.

(f) $2 Al + N_2 \longrightarrow 2 AlN$ (~ 1200 K)

$Al_2O_3 + 3C + N_2 \longrightarrow 2 AlN + 3 CO$ (high temperatures)

$2 ALX_3 + N_2 + 3 H_2 \longrightarrow 2 AlN + 6 HX$ (~ 1600 K)

Decomposition of ammonium hexafluoroaluminate also yields AlN.

(g) $Na_2B_4O_7 + 7 C + 2 N_2 \longrightarrow 4 BN + 7CO + 2 Na$

$B_2O_3 + 2 NH_3 \longrightarrow 2 BN + 3 H_2O$

$B_2O_3 + C + N_2 \longrightarrow 2 BN + 3 CO$

$BCl_3 + NH_3 \longrightarrow BNH_3Cl_3 \longrightarrow BN + 3 HCl$

References

1. H. Nowotny in *Inorganic Chemistry series one,* Vol. 10, Solid State Chemistry (L.E.J. Roberts, ed.) MTP International Rev. Sci., Butterworths, London, 1972.

2. J.J. Ritter and K.C. Frase in *Science of Ceramic Chemical Processing* (L.L. Hench and D.R. Ulrich, eds), John Wiley, New York, 1986.

3. E.K. Storms in *Inorganic Chemistry series One*, Vol. 10, Solid State Chemistry (L.E.J. Roberts, ed). MTP International Rev. Sci., Butterworths, London, 1972.

4. L.L. Hench and D.R. Ulrich (eds), *Science of Ceramic Chemical Processing*, John Wiley, New York, 1986.

5. J.C. Fitzmaurice, A. Hector and I.P. Parkin, *Polyhedron*, **12** (1993) 1295; **13** (1994) 235.

6. M.C. Flowers, N.B.H. Jonathan, A.B. Laurie, A. Morris and G.J. Parker, *J. Mater. Chem.* **2** (1992) 365.

7. M.Y. Chern, D.A. Vennos and F.J. DiSalvo, *J. Solid State Chem.* **96** (1992) 415.

8. D.A. Vennos, M.E. Badding and F.J. DiSalvo, *Inorg. Chem.* **29** (1990) 4059.

9. P.P. Edwards and cowokers, To be published.

10. D.Bem and H-C Zur Loye, *J. Solid State Chem.* **104** (1993) 467.

11. J. Mukherjee in *Chemistry of Advanced Materials* (C.N.R. Rao ed), Blackwell, Oxford, 1992.

12. R. Collongues, J.C. Gilles, A.M. Lejus, M.P. Jorba and D. Michel, *Mater. Res. Bull.*, **2** (1967) 837.

13. S. Ikeda, T. Yagi, N. Ishizawa, N. Mizutani and M. Kato, *J.Solid State Chem.* **73** (188) 52.

14. M. Ohashi, H. Yamamoto, S. Yamanaka and M. Hattori, *Mater. Res. Bull.* **28** (1993) 513.

15. K.H. Jack, *Mater. Res. Bull.* **13** (1978) 1327.

16. P.D. Ramesh and K.J. Rao, *J. Mater. Res.* (1994), in print.

18

Metal Fluorides

Metal fluorides exhibit interesting properties and provide model systems to understand electronic, dielectric and magnetic behaviour of complex solids. Fluorine reacts spontaneously with metals and their compounds to yield fluorides. Many procedures have been employed for the preparation of metallic fluorides and the subject has been reviewed adequately [1–5]. Because of the high reactivity of F_2, HF and other reactive fluorine compounds, special reaction vessels and metallic vacuum lines are employed. Teflon, Kel'f and FEP vessels are generally used although quartz and pyrex vessels can be used for F_2 reactions at moderate temperatures if the formation of HF can be prevented. Reaction of F_2 under high pressure is carried out to produce fluorides, but special precautions and equipment are necessary for the purpose.

Fluorides are prepared by (a) gas phase reactions with F_2 or with fluorides (e.g. MoF_6, SF_4), (b) reaction with HF at atmospheric pressure, (c) reaction with halogen fluorides (e.g. BrF_3, BrF_5, IF_5), (e) reaction with fluorides of Se, Sb and V, (f) reaction with liquid HF and (g) reaction with other fluorinating agents. We give typical examples below:

$$Xe\ (g) + PtF_6\ (g) \longrightarrow XePtF_6\ (s)$$

$$Be\ (OH)_2 + 4\ HF + NiO \longrightarrow NiBeF_4 . 6\ H_2O$$

$$6\ MF + TiO_2 \xrightarrow[\text{Liq . HF}]{} M_2TiF_6 \quad (M = alkali\ metal)$$

$$3\ PdBr_2 + 3\ GeO_2 + 6\ BrF_3 \longrightarrow 3\ PdGeF_6 + 3\ O_2 + 6\ Br_2$$

$$6\ CsBr + 6\ IrBr_4 + 12\ BrF_3 \longrightarrow 6\ CsIrF_6 + 21\ Br_2$$

$$MF + BrF_5 \longrightarrow MBrF_6 \quad (M = alkali\ metal)$$

$$6\ PdF_3 . BrF_3 + 12\ SeF_4 \longrightarrow 6\ (SeF_4)_2\ PdF_4 + Br_2 + 4\ BrF_3$$

$$(SeF_4)_2PdF_4 \xrightarrow{\text{heat}} PdF_2 + SeF_4 + SeF_6$$

Reaction of gaseous HF with metals, oxides, halides etc. to yield fluorides is a route commonly employed by many workers. We illustrate this with a few examples below:

$$Mn, Co \xrightarrow{450\,K} MnF_2, CoF_2$$

$$U \xrightarrow{770\,K} UF_4$$

$$CrCl_3 \xrightarrow{820\,K} CrF_3$$

$$FeCl_3 \xrightarrow{570\,K} FeF_3$$

$$ThO_2 \xrightarrow{870\,K} ThF_4$$

$$Y_2O_3 \xrightarrow{90\,K} YF_3$$

$$CsF + MCl_2 \xrightarrow{870\,K} CsMF_3 \qquad (M = Fe, Co, Ni)$$

$$K_2ReBr_6 \xrightarrow{470\,K} K_2ReF_6$$

Reaction of fluorine (at low pressures) with metals at temperatures around 500–600 K yields fluorides. Examples of fluorides prepared in this manner are: TiF_4, VF_5, CrF_3, MnF_3, MnF_4, MoF_5, MoF_6, NbF_5, AgF, AgF_2, PtF_5, PtF_6.

Reactions of fluorine with oxides, halides and sulfides yields fluorides of many metals as illustrated by the following examples:

V_2O_5 to VOF_3 (745 K); CeO_2 to CeF_4 (770 K);

Tl_2O_3 to TlF_3 (570 K); ZrO_2 to ZrF_4 (795 K),

CdS to CdF_2 (570 K), CuO to CuF_2 (670 K), $AuCl_3$ to AuF_3

(770 K) and PbF_2 to PbF_4 (570 K).

More complex fluorides prepared similarly are exemplified by the following: $CaPbO_3$ to $CaPbF_6$ and $BaFeO_{2.5}$ to $BaFeF_5$. Reactions of F_2 with mixtures of compounds is also carried out to prepare complex fluorides:

$$MnO_2 + 2\,LiCl \xrightarrow{\;720\,K\;} Li_2MnF_6$$

$$Rb_2CO_3 + Bi_2O_3 \xrightarrow{\;720\,K\;} RbBiF_6$$

$$2\,RbF + KF + CuF_2 \xrightarrow{\;520\,K\;} Rb_2KCuF_6$$

SF_4 has been widely used as a fluorinating agent. Thus, WO_3 and MoO_3 give WF_6 and MoF_6 as products on reacting with SF_4. ClF_3 is another good fluorinating agent (e.g. $CoCl_2 \rightarrow CoF_2$, $CeF_3 \rightarrow CeF_4$).

KHF_2 and NH_4HF_2 can be used as fluorinating agents to yield complex fluorides.

$$RuI_3 + 3\,KHF_2 \longrightarrow K_3RuF_6 + 3\,HI$$

Several reactions occur in solid state as well.

$$2\,LiF + CaF_2 + ZrF_4 \longrightarrow ZrCaLi_2F_8$$

$$2\,AF + BF + MF_3 \longrightarrow A_2BMF_6$$
$$(A, B = \text{alkali metal}, M = Ti, V, Cr)$$

$$MF_2 + M'F_3 \longrightarrow MM'F_5$$

$$xMF + xFeF_2 + (1 - x)\,FeF_3 \longrightarrow M_xFeF_3$$

There are many other reactions of relevance to the preparation of fluorides. Some occur at high pressures. Some involve decomposition reactions (of hydrates or fluorides). Reduction reactions with metals, with CO, PF_3 and other reagents as well as thermal treatment are used to prepare fluorides of metals in lower oxidation states ($EuF_3 \rightarrow EuF_2$, $MoF_5 \rightarrow MoF_3$ etc).

There have been several attempts to fluorinate $YBa_2Cu_3O_7$. Many of the results are not reproducible and fluorination in no way improved superconductivity. Solid state synthesis using BaF_2, YF_3 and CuF_2 as well reactions with F_2, NF_3, ClF_3, and ZnF_2 have been carried out [6]. Effect of fluorination of La_2CuO_4 has been compared to that on La_2NiO_4 [7]. Very little fluorine substitution occurs in these two oxides. On the other hand, fluorination of Nd_2CuO_4 yields $Nd_2CuO_{4-x}F_x$ with electron super-conductivity, just as substitution of Nd by Ce^{4+} or Th^{4+} [8].

Reference

1. R.D. Peacock, *Progr. Inorg. Chem.* **2** (1960) 193.
2. A.G. Sharpe, *Adv. Fluorine Chem.* **1** (1960) 29.

3. J.H. Simons (Ed). *Fluorine Chemistry*, Vol. 5, Academic Press, New York, 1964.

4. N. Bartlett, *Prep. Inorg. React.* **2** (1965) 301

5. J. Grannec and L. Lozano in *Inorganic Solid Fluorides* (P. Hagenmuller, ed), Academic Press, New York, 1985.

6. A.R. Armstrong and P.P. Edwards, *Annual Reports C.,* Royal Society of Chemistry, 1991.

7. V. Bhat, C.N.R. Rao and J.M. Honig, *Solid State Commun* **81** (1992) 751.

8. A.C.W.P. James, S.M. Zaharuk and D.W. Murphy, *Nature* **338** (1989) 240.

19

Metal Silicides, Phosphides, Sulfides and Related Materials

Metal silicides are prepared by the direct reaction of the elements. Reduction of a mixture of SiO_2 and the metal oxide by carbothermic reaction provides another route. Reaction of metal oxides with silicon or SiC also yields silicides. Certain silicides can be made electrochemically (e.g. Cr_3Si). Silicides of various element have been reviewed by Nowotny [1].

Nb_3Si is prepared by vapour phase transport. Niobium metal and SiO_2 do not react when heated in vacuum at high temperatures (1370 K). In the presence of traces of H_2, gaseous SiO is formed. Gaseous SiO migrates to niobium to form the silicide.

$$SiO_2 \text{ (s)} + H_2 \rightleftarrows SiO \text{ (g)} + H_2O$$

$$3 \text{ SiO (g)} + 8 \text{ Nb} \longrightarrow Nb_5 Si_3 + 3 \text{ NbO}$$

Another procedure would be to make used of I_2.

$$Nb \text{ (s)} + 2 I_2 \rightleftarrows NbI_4 \text{ (g)}$$

$$11 \text{ NbI}_4 + 3 \text{ SiO}_2 \longrightarrow Nb_5 Si_3 + 22 I_2 + 6 \text{ NbO}$$

Metal phosphides (M_3P to MP_3) are generally prepared by the direct union of elements. Certain phosphides have been prepared by electrolysis (e.g. FeP). Reaction of oxides or halides with PH_3 also yields phosphides.

$$Ga_2O_3 + 2 PH_3 \longrightarrow 2 \text{ GaP} + 3 H_2O$$

$$3 \text{ ZnCl}_2 + PH_3 \longrightarrow Zn_3P_2 + 2 \text{ HCl}$$

Reduction of phosphates by carbon or hydrogen has been employed to prepared phosphides. Reaction of the metal with Ca_3P_2 is another route.

$$Ca_3P_2 + 2\,Ta \longrightarrow 2\,TaP + 3\,Ca$$

Metal phosphides have been reviewed by Corbridge [2]. InP is prepared by starting with an organic indium precursor (e.g., indium alkyls) with PH_3 or t–$BuPH_2$. Similar reactions are employed for the synthesis of GaAs and such compounds (see Sec. 4). Trimethylgallium and trimethylaluminium on reaction with hydrides of group V elements (e.g. PH_3) gives III–V compounds. Another method of preparing GaAs involves the reaction of $AsCl_3$ with Ga. In vapour phase (in the presence of H_2), this reaction directly yields GaAs. Reaction of AsH_3 and gallium in the presence of HCl is also used to prepare GaAs. Organic precursors of aluminium with arsenic have been described [3].

Metal sulfides and other chalcogenides are generally prepared by the direct reaction of the elements. High pressures are employed in the preparation of rare earth sulfides. Flahaut [4] has reviewed the solid state chemistry of metal chalcogenides. Many complex metal sulfides are prepared by the reaction of H_2S with metal oxides. Reaction of oxides with CS_2 vapour is employed for the preparation of sulfides, though not commonly. Sulfides and other chalcogenides are also prepared by using organometallic precursors (see Section 4). For example, ZnTe is obtained by the reaction of the dialkyls of the two elements. Both metal sulfides and phosphides have been prepared by the reaction of anhydrous metal halides with alkali metal pnictides [5] and chalcogenides [6].

The reactive flux method has been useful to synthesize ternary and quarternary transition metal chalcogenides [7]. In this method, a flux of the type A_2Ch_n/Ch (A = alkali metal or Cu, Ch = S, Se/or Te) reacts with one or two transition metals to give compounds of the type $A_xM_yCh_z$ or $A_xM_yM_pCh_z$. Typical examples are $K_nCu_{3-n}NbSe_4$ ($0 < n < 3$), Cu_2MTe_3 (M=Ti,Zr,Th) and $ACuMCh_3$ (A=Na,K; M=Ti,Zr,Th). Open framework structures based on Se_x^{2-} fragments, such as $(Ph_4P)[M(Se_6)_2]$ (M = Ga, In, Tl), have been prepared recently, by employing molten $(Ph_4P)_2\,Se_x$ [8].

References

1. H. Nowotny in *Inorganic Chemistry series one*, Vol. 10, Solid State Chemistry (L.E.J. Roberts ed), MTP International Rev. Sci, Butterworths, London, 1972

2. D.E.C. Corbridge, *Phosphorus*, Elsevier, Amsterdam, 1985

3. R.L. Wells, A.T. McPhail and T.M. Speer, *Eur. J. Solid State Inorg. Chem.* **29** (1992) 63

4. J. Flahaut in *Inorganic Chemistry Series one*, Vol. 10. Solid State Chemistry (L.E.J. Roberts, ed). MTP International Rev. Sci, Butterworths, London, 1972

5. R.E. Treece, G.S. Macala and R.B. Komer, *Chem. Mater.* **4** (1992) 9

6. P.R. Bonneau, R.F. Jarvis and R.B. Kaner, *Inorg. Chem.* **31** (1992) 2127

7. P.M. Keane, Y.J. Lu and J.A. Ibers, *Acc. Chem. Res.*, **24** (1991) 223.

8. S. Dhingra and M.G. Kanatzidis, *Science,* **258** (1992) 258.

20

Nanomaterials

Particles of metals, semiconductors or ceramics having diameters in the range 1–50 nm constitute nanoscale clusters. Physical properties of such clusters correspond neither to those of the free atoms or molecules making up the particle nor to those of bulk solids of the same chemical composition. The clusters are characterized by a large surface area to volume ratio which implies that a large fraction of the atoms reside at the grain boundary. Metastable structures can therefore be generated in nanoclusters and these are different from those of the bulk. Employing such clusters as precursors, one can generate new metastable phases of a given substance.

The science of nanoscale clusters has attracted considerable attention in recent years [1–3]. Several techniques have been developed to prepare clusters of different systems and various tools of characterization have been employed to investigate their properties. Most of the investigations have been carried out on clusters prepared by gas-phase condensation which provides a substrate-free configuration for the nanoscale particles. Clusters grown in liquid and solid media have also been investigated.

The various methods or preparation employed to prepare nanoscale clusters include evaporation in inert-gas atmosphere, laser pyrolysis, sputtering techniques, plasma techniques and chemical methods. In Table 20.1, we list typical materials prepared by inert gas evaporation and sputtering techniques. In Table 20.2 we list typical materials prepared by chemical methods. Nanoparticles of oxide materials can be prepared by the oxidation of fine metal particles, by spray techniques, by precipitation methods (involving the adjustment of reaction conditions, pH etc), or by the sol-gel method.

Nanophase materials are prepared by compacting the nanosized clusters generally under high vacuum. Synthesis of such nanomaterials has been reported in a few systems. The average grain sizes in these materials range form 5 to 25 nm. The properties exhibited by the materials synthesized hitherto are quite different, often markedly improved, compared to the coarsegrained counterparts of the same chemical composition. The atomic arrangements at the interfaces of the nanosized grains are such that different metastable configurations can be induced in them by changing the initial composition. This enables the preparation of a wide range of materials with novel properties.

Table 20.1

Typical Nanoclusters prepared by physical methods[a]

Material	Cluster size (nm)	Method of characterization
(i) By inert-gas evaporation		
CdS	20–250	Optical
CuCl	17–60	Optical
Fe	5–40	Mössbauer
Er	10–70	ESR
Au, Pd	1–10	Microscopy (HREM, STM) Tunnelling Conductance
(ii) By-sputtering		
Al–Cu	10–500	Microstructure
CdS	1.5–2.6	Optical
GaAs	1.6–3.2	Optical
$Co_{0.4}Fe_{0.4}B_{0.2}$	5	Electrical
Ni	1–10	Magnetic
Au, Pd	1–10	Microscopy (HREM, STM)

(a) Clusters of carbon, metals and metal carbides (e.g. M_8C_{12}, M = V, Zr, Hf or Ti) etc. are prepared by laser ablation.

Table 20.2

Typical Nanoclusters prepared by chemical methods

Material	Method	Cluster size (nm)	Method of characterization
$MnFe_2O_4$	Coprecipitation	5–35	Magnetic Mössbauer EXAFS
HgSe	Colloidal preparation	2–10	Optical
PbI_2	Colloidal preparation	1.2–3.6	Optical
Ag	Colloidal preparation	4	Microstructure

Table 20.2 (contd.)

ZrO_2, TiO_2	Gel precipitation, sol-gel	5–20	Structure
Al_2O_3–ZrO_2	Chemical polymerization and precipitation	20–80	Structure
$PbTiO_3$	Sol-gel	12	Ferroelectricity
Nd–Fe–Zr–B	Glass ceramics	20–40	Microstructure
Fe–Co	Pyrolysis of Organometallics	10–50	Magnetic

Nanocomposites [4], as distinct from nanophase materials, consisting of more than one Gibbsian solid phase with at least one dimension in the nanometre range have attracted some attention. These composites lead to monophasic or multiphasic ceramics, glasses or porous materials, with tailored and improved properties. Nanocomposites can be derived from sol-gel, intercalation or entrapment. A large family of microcomposite electroceramic materials, superior in performance to single phase materials, have been prepared [5].

Chemical means of preparing nanoparticles offer many possibilities, the case of metal particles discussed below, being illustrative. Microemulsions and micelles can be employed as the media to produce small particles of 1.5-10 nm diameter with narrow size distribution (e.g. Cds). The sol-gel technique also give small particulates of many oxidic materials. Recently, homogeneous nanoparticles of ZnO and of the oxalate precursor of $YBa_2Cu_3O_7$ have been prepared by coprecipitation in aqueous cores of water-in-oil microemulsions [6]. Nanoparticles of oxide materials are also obtained by nebulized spray pyrolysis (see section 12).

Metal particles: We shall examine the chemical preparation of metal particles in some detail. Chemical preparation of metal colloids initiated by the classic work of Faraday, has remained an art as well as an experimental challenge for many decades. A number of procedures have been employed for the preparation of metal sols. In general, the preparation involves the treatment of a metal salt solution with a suitable reducing agent (e.g., $NaBH_4$, hydroxylamine) in the presence of a protective agent; the latter is to prevent the coagulation of the colloidal particles. The undesired end products are sometimes removed by cellulose dialysis. It is to be noted that the size distribution of the particles depends on the method employed for the preparation.

Several methods have been described for the preparation of gold and platinum sols. Trukevich et al [7] obtained gold colloids in the 7.5–17.5 nm range by the reduction of chloroauric acid using sodium citrate in aqueous medium. They carried out a systematic study of the variation in the size distribution of the particles as a function of both temperature and concentration of the solutions. Sols containing small gold particles (~6nm) have been synthesized by Wilenzick et al [8] who employed phosphorus in ether as a reducing agent. Duff et al [9] have recently suggested a much safer method involving the reduction of $HAuCl_4$ by using tetrakisdihydroxy-methyl-phosphonium chloride in an alkaline medium. This procedure yields particles in the range of ~1 nm.

A variety of reducing agents have been used in the preparation of platinum sols. The citrate reduction of chloroplatinic acid [10, 11] has been found to give Pt particles as small as 2 nm while dimethylamine borane (DMAB) produces particles with a wide size distribution (5–20 nm). Small particles (<3nm) can also be obtained using sodium borohydride as the reducing agent [11] in presence of polyvinyl pyrrolidone (PVP). In the absence of such a protective agent, the reaction mixture yields strand-like aggregates composed of ~6 nm diameter links. Synthesis using hydroxyl amine hydrochloride in the presence of PVP at 277K gives unsintered Pt particles of mean diameter ~2 nm. In addition, colloidal suspensions of Pt-group metals and bimetals [12–15] have been prepared by the reduction of chloroplatinic acid using methanol in the presence of PVP. In some cases, the reducing power of the solution has been increased by adding glucose [12].

A more general procedure has been suggested by Dye and coworkers [16, 17] for the preparation of sols of a number of metals. Alkali metals solubilized in aprotic solvents such as dimethylether or tetrahydrofuran along with a suitable cation complexant (such as a crown ether or a cryptand) has been employed by these workers as an attractive reducing medium. Homogeneous reduction of various metal salts with alkalides or electrides in dimethyl ether or THF at 223K or below produces 2–15 nm diameter particles of metals and alloys. The reactions are rapid, complete, and applicable to a wide range of elements from Ti to Te. Synthesis involving highly reactive metals (3d metals) requires an inert atmosphere.

Large metal particles (micrometer dimensions) have also been prepared by chemical methods involving the reduction of metal salts [18, 19]. In the polyol process, the metal salts are reduced by glycerol, ethylene glycol etc. Monodispersed metal particles of different shapes have been obtained in this manner.

References

1. R.P. Andres, R.S. Averback and W.L. Brown, *J. Mater Res.* **4** (1989) 704.
2. D. Chakravorty and A.K. Giri in *Chemistry of Advanced Materials* (C.N.R. Rao, ed), Blackwell, Oxford, 1992; R.W. Siegel in *Advances in Materials and their Applications* (P. Rama Rao, ed, Wiley-Eastern, New Delhi, 1993).

3. P. Jena, B.K. Rao and S.N. Khanna (eds). *Physics and Chemistry of Small Clusters*, Plenum Press, New York, 1986; also *Physics and Chemistry of Finite Systems: From Clusters to Crystals*, Kluwer Publishers, Dordrecht, 1992.

4. S. Komarneni, *J. Mater. Chem.* **2** (1992) 1219.

5. R.E. Newnham, S.E. McKinstry and H. Ikawa, *Mater. Res. Soc. Symp. Proc.* **175** (1990) 161.

6. P. Kumar, P. Pillai, S.R. Bates and D.O. Shah, *Mater. Lett.*, **16** (1993) 68; Also, S. Hingorani et al, *Mat. Res. Bull 28* (1993) 1303.

7. J.S. Turkevich and P.S. Stevenson, J. Hiller, *Discuss. Faraday Soc.* **11** (1951) 55.

8. R.M. Wilenzick, D.C. Russell, R.H. Morris and S.W. Marshall, *J. Chem. Phys.*, **47** (1967) 533.

9. D.G. Duff, A. Baiker and P.P. Edwards, *J. Chem. Soc. Chem. Commun.* (1992) 96.

10. D.N. Furlong, A. Launikonis and W.H.F. Sasse and J.V. Sanders, *J. Chem. Soc., Faraday Trans.* 1, **80** (1984) 571.

11. P.R. Van Rheenen, M.J. McKelvy and W.S. Glausinger, *J. Solid State Chem.*, **67** (1987) 151.

12. D. Richard, J.W. Couves and J.M. Thomas, *Faraday Discuss. Chem. Soc.* **92** (1991) 109.

13. M. Harada, K. Asakura and N. Toshime, *J. Phys. Chem.* **97** (1993) 5103.

14. M. Harada, K. Asakura, Y. Ueki and N. Toshime, *J. Phys. Chem.* **96** (1992) 9730.

15. N. Toshima, M. Harada, U. Yamazaki and K. Asakura, *J. Phys. Chem.*, **96** (1992) 9927.

16. K.-L. Tsai and J.L. Dye, *J. Am. Chem. Soc.* **113** (1991) 1650.

17. K.-L. Tsai and J.L. Dye, *Chem. Mater.* **5** (1993) 540.

18. F. Fievet, J.P. Lagier and F. Figlarz, *MRS Bulletin* **14** (1989) 29.

19. E. Matijevic, *Faraday Discuss. 92* (1991) 229; *Acc. Chem. Res.* **14** (1981) 22.

SUBJECT INDEX